Student Support Materials for

AQA

AS Physics

Unit 1: Particles, Quantum Phenomena and Electricity

Author: Dave Kelly

William Collins's dream of knowledge for all began with the publication of his first book in 1819. A self-educated mill worker, he not only enriched millions of lives, but also founded a flourishing publishing house. Today, staying true to this spirit, Collins books are packed with inspiration, innovation and practical expertise. They place you at the centre of a world of possibility and give you exactly what you need to explore it.

Collins. Freedom to teach.

Published by Collins
An imprint of HarperCollinsPublishers
77-85 Fulham Palace Road
Hammersmith
London
W6 8JB

Browse the complete Collins catalogue at
www.collinseducation.com

10 9 8 7 6 5 4 3 2 1

ISBN-13 978-0-00-734383-6

Dave Kelly asserts his moral right to be identified as the author of this work.

British Library Cataloguing in Publication Data. A Catalogue record for this publication is available from the British Library.

Thanks to John Avison and Stuart Jones for their contributions to the previous editions.

Commissioned and Project Managed by Letitia Luff
Edited and proofread by Jane Roth
Typesetting by Hedgehog Publishing
Cover design by Angela English
Index by Jane Henley
Production by Leonie Kellman
Printed and bound in Hong Kong by Printing Express

Mixed Sources
Product group from well-managed
forests and other controlled sources
www.fsc.org Cert no. SW-COC-1806
© 1996 Forest Stewardship Council

FSC is a non-profit international organisation established to promote the responsible management of the world's forests. Products carrying the FSC label are independently certified to assure consumers that they come from forests that are managed to meet the social, economic and ecological needs of present and future generations.

Find out more about HarperCollins and the environment at
www.harpercollins.co.uk/green

Contents

3.1.1 **Particles and radiation**

Constituents of the atom; stability of the nucleus

Essential Notes

Electrons are thought to be fundamental particles; there is no evidence that the electron can be broken down into any other particles.

The electron

The electron was first identified during experiments using electrical discharge tubes (Fig 1). When the voltage is turned on, the screen at the end of the tube emits a glow. The glow was said to be caused by 'cathode rays'. When the rays hit the screen, their energy is converted into light. This energy conversion, known as **fluorescence**, is aided by coating the inside of the screen with a phosphor, such as zinc sulphide.

Fig 1
An electrical discharge tube

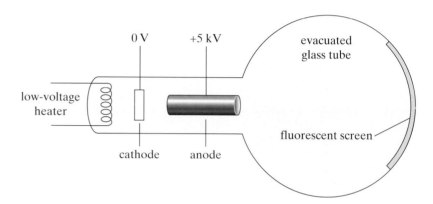

Examiners' Notes

The **charge–mass ratio** of a particle is its charge in coulombs divided by its mass in kilograms. This is also called its **specific charge** and has units of $C\ kg^{-1}$.

In 1897, J. J. Thomson discovered that he could deflect the rays using electric or magnetic fields (Fig 2). He balanced the two deflections so that the rays moved in a straight line. This allowed him to calculate the **charge–mass ratio** for the particles making up the rays. He concluded that cathode rays were tiny, negatively charged particles, now called **electrons**.

Fig 2
Electrostatic and magnetic deflection: (a) cathode rays curve towards the positive plate, showing they carry a negative charge; (b) a magnetic field makes the cathode rays move in a circular path

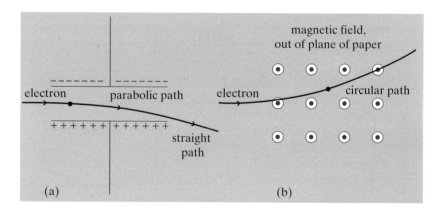

Examiners' Notes

The force on the electrons acts from the negative cathode to the positive anode: the reverse of the direction of the electric field. The force on the electrons in a magnetic field is at right angles to the field and the motion of the electrons.

Electrons are emitted from the negative electrode, or **cathode**. More electrons are emitted if the cathode is heated. This effect is known as **thermionic emission**. After they are emitted the electrons accelerate towards the anode, finally hitting the screen, where they cause fluorescence.

Thomson realised that the electrons were torn away from atoms in the cathode's surface by the electric field. He suggested that atoms were composed of many electrons, moving in various orbits inside a positively charged cloud. This model of atomic structure is often referred to as the 'plum pudding' model of the atom (Fig 3).

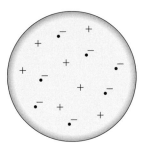

Fig 3
Simple picture of 'plum pudding' model of atom

Properties of the electron

Mass of the electron $m_e = 9.1 \times 10^{-31}$ kg

Charge of the electron $e = -1.6 \times 10^{-19}$ C

Charge–mass ratio $\dfrac{e}{m_e} = 1.76 \times 10^{11}$ C kg^{-1}

Essential Notes

The charge on an electron is very small. One coulomb is the amount of charge that flows past a point when 1 ampere of current flows for one second. If this was all carried by electrons, there would be 6.25×10^{18} electrons flowing past each second.

The nuclear atom

The plum pudding model of the atom had to be abandoned following Rutherford's scattering experiments. These showed that atoms were almost all empty space but with a very small, very dense positively charged **nucleus** (see p. 7). In our present model, all the positive charge of the atom and nearly all the mass are located in the central nucleus. Tiny, negatively charged electrons orbit this nucleus, rather like planets orbiting the Sun (Fig 4).

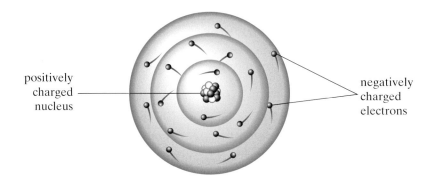

positively charged nucleus

negatively charged electrons

Fig 4
Nuclear atom

Definition

Protons are the particles that carry the positive charge in the nucleus.

In a neutral atom, that is an atom with no net charge, the number of protons in the nucleus is balanced by the number of electrons orbiting the nucleus. A hydrogen atom has one proton and one electron. Helium has two protons and two electrons, and so on through the Periodic Table, until the heaviest naturally occurring element, uranium, which has 92 protons and 92 electrons.

Nuclear stability

What holds the nucleus together? Positive charges repel each other and at such short distances the electrostatic forces pushing the nucleus apart are very large. Another force acts inside the nucleus, known as the **strong nuclear force** or **strong interaction**. The strong nuclear force has a very short range; it has no effect at separations greater than about 5 fm (5×10^{-15} m). The strong interaction is an attractive force until the separation is less than 1 fm, when the force becomes strongly repulsive (Fig 5). The overall effect of the strong interaction is to pull the nucleus together, but the repulsive action prevents it from collapsing to a point.

Essential Notes

The **femtometre (fm)** is a useful unit in nuclear physics.

1 fm = 10^{-15} m

Essential Notes

At distances of less than about 2 fm, the strong nuclear attraction between two protons is larger than the electrostatic repulsion (Fig 5) so the nucleus is held together.

Fig 5
Force–distance graph for proton–proton pair (strong nuclear force plus electrostatic force)

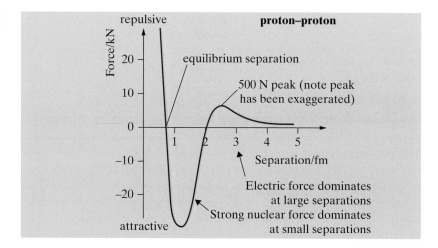

For large nuclei there is a problem. The strong nuclear force acts over a much shorter range than electrostatic repulsion. It isn't possible to get all the protons close enough together for the strong nuclear force to overcome the electrostatic repulsion. There has to be some other particle in the nucleus that helps to glue it all together and keep it stable. This is the **neutron**, discovered by James Chadwick in 1932.

Fig 6
Force–distance graph for strong nuclear force without electrostatic repulsion

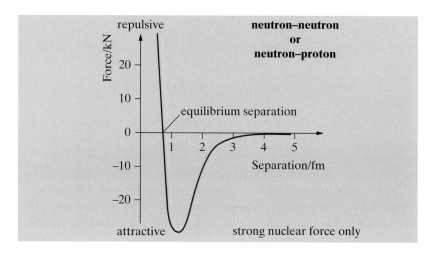

Definition

*The **neutron** is a particle with a mass almost identical to that of the proton, but with no electric charge.*

The neutron exerts a strong nuclear attraction on protons and on other neutrons. Protons and neutrons are the only particles in the nucleus. They are referred to as **nucleons**. The strong nuclear force acts between *any* pair of nucleons, whether that is two protons, two neutrons, or a proton and a neutron. Electrostatic repulsion acts only between protons.

The properties of protons, neutrons and electrons are summarised in Table 1.

	Proton	Neutron	Electron
symbol	p	n	e^-
charge (C)	$+1.602 \times 10^{-19}$	0	-1.602×10^{-19}
mass (kg)	1.6726×10^{-27}	1.6749×10^{-27}	9.109×10^{-31}
specific charge or charge–mass ratio (C kg^{-1})	9.58×10^6	0	1.78×10^{11}
charge (relative to proton)	1	0	-1
mass (relative to proton)	1	$1.0014 \approx 1$	5.45×10^{-4}

Table 1
Comparison of protons, neutrons and electrons

Evidence for the existence of the nucleus

Geiger and Marsden, working in Ernest Rutherford's laboratories at Manchester University in 1909, studied the scattering of alpha particles as they passed through a thin piece of gold foil (Fig 7). **Alpha particles** are relatively massive, positively charged particles emitted by some radioactive materials.

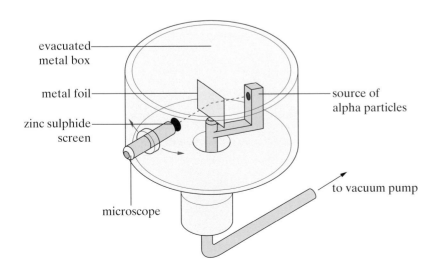

Essential Notes

Alpha particles are two protons and two neutrons bound tightly together, the same configuration as a helium nucleus. An alpha particle therefore carries a positive charge which is twice the size of the charge on the electron. The mass of an alpha particle is about 8000 times the electron's mass.

Fig 7
Geiger and Marsden's apparatus

Geiger and Marsden used a scintillator to detect the alpha particles. (A scintillator is a zinc sulphide screen which emits light whenever an alpha particle strikes it.) The scintillator was observed through a small microscope in a darkened room.

Geiger and Marsden expected to see the alpha particles deflected by small angles. Rutherford suggested that they move the detector in front of the foil, to see if any of the alpha particles were bounced from the surface of the foil. Amazingly, some were; about 1 in every 8000 alphas was 'reflected', or scattered, through an angle of more than 90°.

Essential Notes

Note that the alpha particles bombard the foil with a lot of energy – turning them round and sending them back requires a very strong force!

As an alpha particle travelling at around 10 000 km s^{-1} could not be bounced back by a positively charged cloud with tiny electrons embedded in it, Rutherford concluded that almost all the mass of the atom must be gathered together in one small volume, which he called the nucleus. He suggested that the electrons carry all the negative charge and that they orbit the nucleus through empty space a relatively long way from the nucleus. Most of the alpha particles passed through the gold foil with small or zero deflections because they were too far away from the nucleus to be affected by it. Very occasionally an alpha particle would pass so close to the nucleus that it would be repelled by its positive charge and suffer a large deflection (Fig 8).

Fig 8
Rutherford scattering, showing several paths

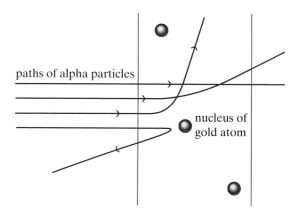

paths of alpha particles

nucleus of gold atom

Analysis of the results allowed Rutherford to work out the scale of the nuclear atom. Whereas the radius of the atom was about 10^{-10} m, the nucleus was only about 10^{-15} m across.

Essential Notes

Remember that density, ρ, is defined as the mass in a unit volume:

$$\rho = \frac{m}{V}$$

and is measured in kg m^{-3}

Example

An atom of gold has 79 electrons orbiting its nucleus. The mass of a gold atom is 3.27×10^{-25} kg. The radius of a gold atom is 1.44×10^{-10} m. The radius of a gold nucleus is 6.5×10^{-15} m. Find the average density of an atom of gold and the density of a gold nucleus.

Answer

If we assume that the atom is spherical,

$$\text{Volume} = \frac{4}{3}\pi r^3$$

$$\text{Density} = \frac{\text{mass}}{\text{volume}}$$

So the average atomic density is

$$\frac{3.27 \times 10^{-25}\,\text{kg}}{\frac{4}{3}\pi(1.44 \times 10^{-10}\,\text{m})^3} = 2.6 \times 10^4\,\text{kg}\,\text{m}^{-3}$$

For the nucleus, first subtract the mass of 79 electrons.

Mass of nucleus $= 3.27 \times 10^{-25}\,\text{kg} - 79 \times 9.11 \times 10^{-31}\,\text{kg}$
$= 3.27 \times 10^{-25}\,\text{kg}$

Density of the nucleus $= \dfrac{3.27 \times 10^{-25}}{\frac{4}{3}\pi(6.5 \times 10^{-15})^3}$

$= 2.84 \times 10^{17}\,\text{kg}\,\text{m}^{-3}$

For gold, the density of the nucleus is around 10^{13} times higher than the average atomic density.

Essential Notes

Note that subtracting the mass of the electrons makes no difference to the calculation. The mass of the electrons is only about 0.02% of the total mass of the atom.

All nuclei have approximately the same density, around $2 \times 10^{17}\,\text{kg}\,\text{m}^{-3}$. A matchbox full of nuclear matter would have a mass of around 8 billion tonnes, $8 \times 10^{12}\,\text{kg}$.

Nuclear composition

The simplest atom is hydrogen. It has one proton in its nucleus and no neutrons. It has only one electron orbiting the nucleus (Fig 9). Helium has two protons and two neutrons in its nucleus, with two electrons in orbit around it.

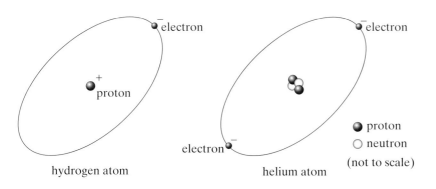

hydrogen atom   helium atom

(not to scale)

Fig 9
The hydrogen and helium atoms

Definition

*The **proton number**, or **atomic number**, is the number of protons in the nucleus and is given the symbol Z.*

The atomic number of an atom is also the number of electrons in the neutral atom. This determines the chemical properties of the atom. The atomic number is used to place elements in the Periodic Table.

Examiners' Notes

The mass number A is always a whole number.

Definitions

*The **nucleon number** A is the total number of nucleons in the nucleus of an atom. It is also known as the **mass number**.*

*The **neutron number** N is the number of neutrons in the nucleus.*

The nucleon number is the number of protons plus the number of neutrons, so

$$A = Z + N$$

The nuclear composition of an atom can be described using symbols. The most common form of carbon has six protons and six neutrons in its nucleus. It can be written as $^{12}_{6}\text{C}$.

The upper number is the nucleon number and the lower number is the atomic number. In general an element X with an atomic number Z and an atomic mass number A is written as $^{A}_{Z}\text{X}$. Using this system the hydrogen nucleus is represented as $^{1}_{1}\text{H}$, and helium as $^{4}_{2}\text{He}$. The first five elements in the Periodic Table are listed in Table 2.

Table 2
The first five elements in the Periodic Table

Element	Symbol	Atomic number, Z	Neutron number, N	Atomic mass number, A
hydrogen	H	1	0	1
helium	He	2	2	4
lithium	Li	3	4	7
beryllium	Be	4	5	9
boron	B	5	6	11

Isotopes

Hydrogen usually has one proton, only, in its nucleus, but some hydrogen atoms have one or two neutrons as well (Fig 10). The different atoms are referred to as **isotopes**.

Definition

Isotopes are forms of an element with the same proton number but with a different number of neutrons.

Fig 10
Isotopes of hydrogen. The extra neutrons do not affect hydrogen's chemical behaviour; for example, all three isotopes combine with oxygen to make water

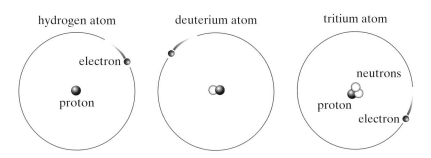

The different isotopes of an element have identical chemical behaviour because their atoms have the same number of electrons. Isotopes also have the same number of protons in their nucleus. The difference is simply in the number of neutrons. This makes some isotopes heavier than others. The isotope of carbon that has six protons and six neutrons in its nucleus, $^{12}_{6}C$, is referred to as carbon-12 (Table 3). Carbon-13 has six protons and seven neutrons; carbon-14 has six protons and eight neutrons.

Isotope	Atomic number, Z	Number of electrons	Neutron number, N	Mass number, A	% abundance
carbon-12	6	6	6	12	98.89
carbon-13	6	6	7	13	1.11
carbon-14	6	6	8	14	< 0.001%

Table 3
Isotopes of carbon

Example

Find the specific charge of the nucleus of an atom of carbon, $^{12}_{6}C$.

Answer

A nucleus of $^{12}_{6}C$ has 6 protons and 6 neutrons. To find the mass of the nucleus add together the masses of the constituents.

Mass of 6 protons $= 6 \times 1.6726 \times 10^{-27}$ kg
Mass of 6 neutrons $= 6 \times 1.6749 \times 10^{-27}$ kg
Total mass $= 2.008 \times 10^{-26}$ kg

The charge on the nucleus is just due to the protons.
Total charge $= 6 \times 1.602 \times 10^{-19}$ C $= 9.612 \times 10^{-19}$ C

$$\text{Specific charge} = \text{charge–mass ratio} = \frac{9.612 \times 10^{-19}\,\text{C}}{2.008 \times 10^{-26}\,\text{kg}}$$

$$= 4.787 \times 10^{7}\,\text{C kg}^{-1}$$

Particles, antiparticles and photons

Antimatter

The British physicist Paul Dirac predicted the existence of a particle with exactly the same mass as the electron, but with a positive charge. Indeed he suggested that all particles must have such **antiparticles**.

Definition

*An **antiparticle** is a 'mirror image' of a particle, of identical mass but opposite charge.*

The first antiparticle was discovered in 1932 by Anderson, who was observing tracks in a cloud chamber made by cosmic rays (high-energy particles from space). He used a strong magnetic field to curve the paths of high-energy electrons. Some tracks seemed identical to the electron tracks but curved in the opposite direction. These were tracks of an antielectron, now known as a **positron**. This is an example of **antimatter**.

Definitions

The **positron** is the electron's antiparticle.

The electron is written $_{-1}^{0}e$ and the positron is written $_{+1}^{0}e$.

Essential Notes

Annihilation is an example of mass-energy equivalence, as predicted by Einstein's Special Theory of Relativity. The total mass–energy in any system is conserved, but energy and mass may be converted from one to the other. Conversion of mass to energy powers radioactivity and nuclear fission.

When a particle meets its antiparticle, the particles are drawn together by electrostatic attraction until they annihilate each other (Fig 11).

Definition

Annihilation is the conversion of the mass of a particle and its antiparticle to a pair of photons of electromagnetic radiation.

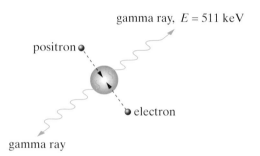

gamma ray, $E = 511\ \text{keV}$

positron

electron

gamma ray

Fig 11
Positron annihilation. When a positron and an electron meet, they annihilate each other. Two identical gamma rays of energy 511 keV are emitted in opposite directions

Annihilation is the conversion of matter to energy. The opposite process, where matter is 'created' from energy, is called **pair production**.

Essential Notes

The joule (J) is the SI unit of energy, but it is too large for measuring energy on an atomic scale. We use the **electron volt** (eV), a much smaller unit.

$1\ \text{eV} = 1.602 \times 10^{-19}\ \text{J}$

1 eV is the energy change when the charge on one electron moves through a potential difference of 1 V.

1 keV (kilo electron volt) = 1000 eV

1 MeV (mega electron volt) = 1 000 000 eV

1 GeV (giga electron volt) = 10^9 eV

1 TeV (tera electron volt) = 10^{12} eV

Definition

Pair production is the process in which a photon of electromagnetic energy is converted to a pair of particles.

In pair production, there are always two particles created; one is a conventional particle and the other is its antiparticle. This satisfies the conservation of charge, since before the event there is only a photon of radiation, which carries no charge. After the pair production there are always two particles of opposite charge, making the total charge zero.

A gamma ray has to have a minimum energy of 1.02 MeV before it can create an electron–positron pair. This is because the mass of the pair has an energy equivalent to 1.02 MeV. If the photon has more energy than this, the surplus energy appears as kinetic energy carried by the positron and electron.

In 1955 the first antinucleon was discovered. Protons were accelerated to an energy of up to 6 MeV and collided into other protons in a stationary target. The two protons (p) collided and produced **antiprotons** ($\bar{\text{p}}$) by the reaction:

$$\text{p} + \text{p} \rightarrow \text{p} + \text{p} + \text{p} + \bar{\text{p}}$$

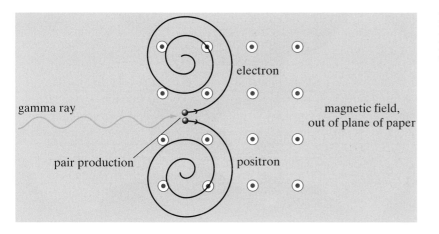

By colliding two protons together we have produced an extra proton and an antiproton. The extra mass needed to create the proton–antiproton pair has come from the kinetic energy of the initial protons.

A year later the **antineutron** ($\bar{n}$) was produced by using antiprotons to collide with protons:

$$\bar{p} + p \rightarrow n + \bar{n}$$

Particles cannot be created unless there is enough energy. The minimum energy needed to create a particle is known as its **rest energy** E_0, which depends on its **rest mass**, m_0. The rest energy is calculated from the rest mass using the formula, $E_0 = m_0 c^2$, where c = the speed of light = 3×10^8 m s^{-1}.

Neutrinos

Neutrinos are probably the most numerous particles in the Universe. They outnumber the protons and neutrons of ordinary matter by a factor of 10^9. Neutrinos created at the time of the Big Bang still permeate the Universe, about 100 or so of them in each cubic centimetre of space. Neutrinos are also emitted by radioactive nuclei and from nuclear reactions. The Earth is bathed in neutrinos from the Sun. Every second about 60 thousand million solar neutrinos pass through every square centimetre of the Earth's surface.

Despite this, neutrinos and antineutrinos are extremely difficult to detect. It wasn't until 1956 that the neutrino was observed experimentally, 26 years after its existence was predicted. Neutrinos are not charged, so they interact with other matter very weakly. Experiments to find neutrinos often use large tanks of water, usually placed deep underground, surrounded by sensitive light detectors looking for the occasional flash of light that signifies that a neutrino has interacted with a neutron, or an antineutrino with a proton.

The neutrino was first predicted by Wolfgang Pauli in 1930. At the time physicists were trying to understand beta radiation – the emission of **beta particles** (fast-moving electrons) from the nuclei of some radioactive atoms. Unlike alpha particles, which are emitted with a well-defined

Examiners' Notes

An antiparticle is denoted by a horizontal line above the symbol for the particle, e.g. $\bar{p}$ denotes an antiproton, and $\bar{n}$ denotes an antineutron. The exception to this is the positron (the electron's antiparticle), which is normally written as e$^+$.

Essential Notes

The rest mass of a particle is the mass when measured in a frame of reference where the mass is stationary. Einstein's Special Theory of Relativity describes how the mass of an object increases as it moves faster, but you don't have to worry about this in Unit 1.

Essential Notes

An antiparticle has the same mass as it conventional 'twin' but opposite charge. What about the antineutron, which has no charge? An antineutron has magnetic properties which are opposite to those of the neutron. It is more difficult to specify the difference between a neutrino and an antineutrino.

energy, beta particles are emitted with a range of energies (Fig 13). This seemed to contravene the principle of conservation of energy. If a certain amount of energy is transferred by each radioactive decay, why did the emitted beta particle have a range of possible energies? Pauli suggested that another particle, the neutrino, is also emitted in beta decay. The neutrino carries away the balance of the energy, so that the total energy of the decay is always constant.

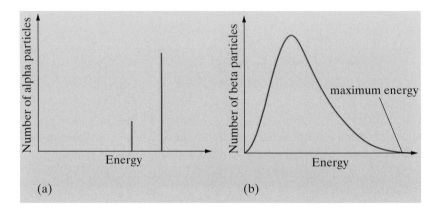

Fig 13
(a) Typical alpha spectrum
(b) Typical beta spectrum

Essential Notes

Alpha particles tend to be emitted from large, unstable nuclei. An alpha particle is two protons and two neutrons bound together (just as in the helium nucleus, $^4_2$He). It escapes from the large (parent) nucleus, leaving a different (daughter) nucleus behind. For example when an isotope of uranium emits an alpha particle, leaving behind an isotope of thorium:

$$^{238}_{92}U \rightarrow {}^{234}_{90}Th + {}^4_2He$$

This is sometimes written as:

$$^{238}_{92}U \rightarrow {}^{234}_{90}Th + \alpha$$

The neutrino is represented by the symbol ν_e and the **antineutrino** is represented by the symbol $\bar{\nu}_e$. The subscript 'e' stands for 'electron'; these neutrinos are more properly referred to as electron neutrinos, because other types of neutrinos exist.

Beta particles are electrons emitted when a neutron decays into a proton and an electron. The proton stays inside the nucleus but the electron is emitted at high speed, together with an antineutrino:

$$^1_0n \rightarrow {}^1_1p + {}^{\ \ 0}_{-1}e + \bar{\nu}_e$$

Some radioactive decays emit a positive beta particle, or positron. This involves the decay of a proton and can be written

$$^1_1p \rightarrow {}^1_0n + {}^0_1e + \nu_e$$

The neutrino is a fundamental particle which carries no charge. For some years it was believed to have zero mass, but experiments now suggest that it has a small mass, much less than that of an electron.

Definition

*The **neutrino** is a fundamental particle with no charge. It has a very small mass. It interacts with other matter very weakly.*

Table 4 summarises the properties of the fundamental particles and their antiparticles.

Particle	Mass/kg	Charge/C	Rest energy/MeV
electron	9.109×10^{-31}	-1.602×10^{-19}	0.511
positron	9.109×10^{-31}	$+1.602 \times 10^{-19}$	0.511
proton	1.6726×10^{-27}	$+1.602 \times 10^{-19}$	938
antiproton	1.6726×10^{-27}	-1.602×10^{-19}	938
neutron	1.6749×10^{-27}	0	939
antineutron	1.6749×10^{-27}	0	939
neutrino	unknown, probably non-zero	0	unknown, probably non-zero
antineutrino	unknown, probably non-zero	0	unknown, probably non-zero

Table 4
Mass, charge and rest energy of particles and antiparticles

Photon model of electromagnetic radiation

The wave theory of light is extremely successful in explaining certain phenomena, such as diffraction and refraction. However, there are some phenomena, such as black-body radiation and the photoelectric effect (see later), which cannot be explained by assuming that electromagnetic radiation is composed of waves.

All objects emit radiation because of their thermal energy. The spectrum of radiation that is emitted depends on the surface temperature of the object and on the type of surface. Objects which are ideal radiators of energy are known as **black-body radiators**.

Essential Notes

A perfect black body is an object that absorbs all the radiation that falls on it and reflects none. If this body is in thermal equilibrium with its surroundings, then it must emit radiation at the same rate as it absorbs it. A perfect black body is therefore an ideal radiator.

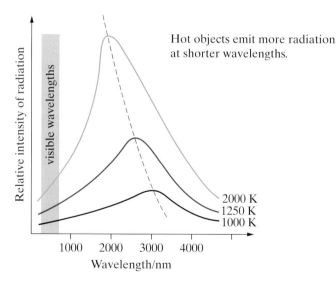

Hot objects emit more radiation at shorter wavelengths.

2000 K
1250 K
1000 K

Fig 14
Black-body radiation curves

The wave theory of radiation could not explain the shape of the black-body radiation curves (Fig 14); in fact the wave theory predicted that at high temperatures infinite amounts of energy would be emitted at short wavelengths (an impossible situation which became known as the 'ultraviolet catastrophe'). In 1900 Max Planck was able to explain the black-body curves by suggesting that the energy was emitted intermittently in packets, called **quanta** (singular 'quantum') of energy. A radiating body may emit an integral number of these packets of energy, say one, two,

Examiners' Notes

$E = hf$ is the energy of a photon of light of frequency f. Since $c = f\lambda$, where c is the speed of light and λ is the wavelength, the equation can be written as $E = hc/\lambda$ which is often useful.

three, etc., but cannot emit any fractional amount. The amount of energy, E, carried by each quantum depends on the frequency, f, of the oscillations that are causing the radiation: $E \propto f$, or

$$E = hf$$

where h is the Planck constant, which has a value of 6.63×10^{-34} J s. The packets or quanta of electromagnetic energy are known as **photons**.

Example

A sodium vapour light emits 30 W of light energy. How many photons are emitted per second? (The wavelength of sodium light is 5.88×10^{-7} m.)

Answer

The frequency of sodium light is given by

$$f = \frac{c}{\lambda} = \frac{3 \times 10^8 \, \text{m s}^{-1}}{5.88 \times 10^{-7} \, \text{m}} = 5.10 \times 10^{14} \, \text{Hz}$$

The energy of one photon of light is

$$E = hf = 6.63 \times 10^{-34} \, \text{J s} \times 5.10 \times 10^{14} \, \text{s}^{-1} = 3.38 \times 10^{-19} \, \text{J}$$

Since the sodium light transfers 30 joules per second, the number of photons per second is

$$\frac{30}{3.38 \times 10^{-19}} = 8.9 \times 10^{19} \, \text{s}^{-1}$$

Particle interactions

Current theories suggest that there are only four types of interaction between particles.

- **Gravity**. The gravitational force has an infinite range and acts on all particles. Although on the scale of the Universe it is the most important of all the fundamental interactions, on an atomic scale it has negligible influence. This is because gravity is the weakest of all the fundamental forces.

- **Electromagnetic force**. The electromagnetic force acts between all charged particles. Because the electromagnetic force holds atoms and molecules together, it is responsible for almost everything that happens to us. Forces such as friction, buoyancy and contact forces are all electromagnetic in origin.

- **Weak interaction**. The weak interaction acts between all particles but over a very short range, about 10^{-18} m. Over this range it is much stronger than gravitation, 10^{33} times as strong in fact. The weak interaction is responsible for radioactive decay.

- **Strong interaction**. The strong interaction, or strong nuclear force, holds nuclei together. It acts between **hadrons** (see later), such as neutrons and protons. It is a short-range force, acting over the nuclear distance scale of around 10^{-15} m. The strong interaction is not felt by **leptons** (see later), such as electrons.

Exchange particles

The Japanese physicist Hideki Yukawa suggested that when two particles A and B exert a force on each other, a **virtual particle** is created. This virtual particle is exchanged between particles A and B and affects their motion.

> **Definition**
>
> An *exchange particle* is a virtual particle, which may exist for only a short time, and is the mediator of a force.

The idea of a force being carried by an exchange particle can be pictured by considering two people on skates (Fig 15). If one of them throws a heavy ball to the other one, both the skaters' motions will be affected; in fact they will be repelled from each other. We have to stretch the analogy a bit to understand attraction, but if you imagine a boomerang being thrown rather than a ball, then the two skaters will be drawn together.

(a)

(b)

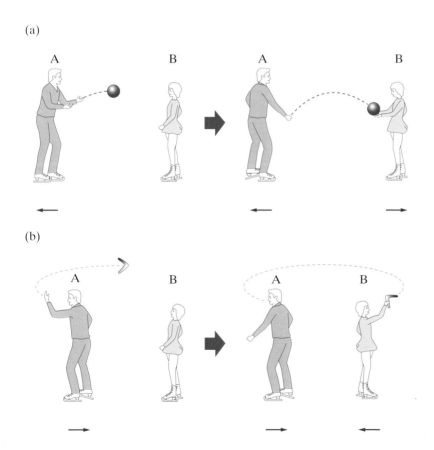

Fig 15
Interactions that can occur when two ice skaters exchange (a) a heavy object, (b) a boomerang

The exchange particles which are transferred between fundamental particles are known as **gauge bosons**. Each fundamental force has its own boson or bosons.

The electromagnetic force

The electromagnetic force is carried between charged particles by the photon, γ. When two charged particles exert a force on each other, a virtual photon is exchanged between them. The photon is a massless, chargeless particle. We can use a **Feynman diagram** to represent what happens (Fig 16).

Fig 16
Feynman diagram showing two electrons feeling the electric force as a result of exchange of a photon

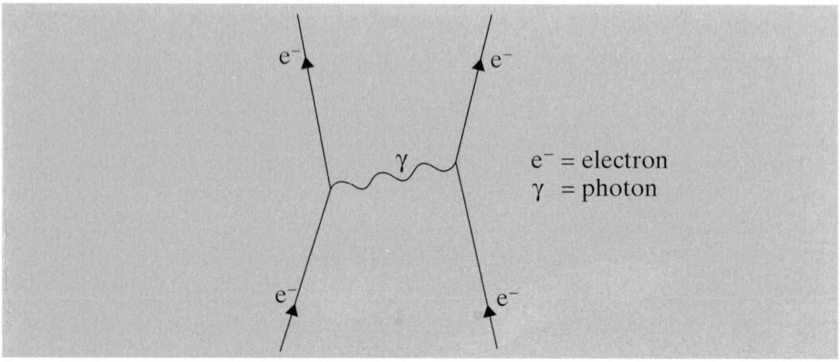

Essential Notes

The Feynman diagram represents the interaction between the particles. The angles of the particle paths are not significant, but time is usually shown going up the page.

Essential Notes

The Uncertainty Principle is one of the key ideas of quantum physics. Although it seems at odds with common sense, the theory is in excellent agreement with experimental results. It means that what we used to imagine as a vacuum is not composed of nothing at all; rather there is constant creation and destruction of virtual particle–antiparticle pairs.

e^- = electron
γ = photon

The strong interaction

It was the strong interaction, acting between nucleons, that Yukawa was working on when he proposed the idea of exchange particles. He suggested that these exchange particles could be travelling at close to the speed of light across the nucleus. But where does the energy to create these particles come from? According to Heisenberg's Uncertainty Principle, particles of energy ΔE can be created for a time Δt, provided that the product $\Delta E \times \Delta t$ does not exceed a certain value h, the Planck constant, which is a very small number, 6.626×10^{-34} J s. The Uncertainty Principle allows particles to appear for a short time before being annihilated again, provided that $\Delta E \times \Delta t < h$.

An exchange particle moving at close to the speed of light has to exist for about 10^{-23} s if it is to have time to travel across the nucleus. This enabled Yukawa to predict a maximum mass for the particle. The particle was discovered in 1947 and was known as a **pi meson** or **pion** (Fig 17).

Fig 17
Feynman diagram showing pion exchange

Essential Notes

Particles are often denoted by a Greek letter such as π, Δ or Σ. A superscript is used to denote charge, e.g. π^+ carries a single positive charge, π^- carries a single negative charge and π^0 is uncharged.

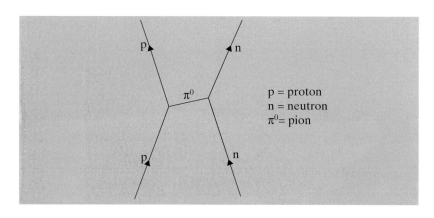

p = proton
n = neutron
π^0 = pion

At a deeper level the strong interaction is mediated by gauge bosons called **gluons** that pass between **quarks** (see later). The pion is simply a vehicle carrying gluons between hadrons. There are eight different gluons, none of which has ever been detected as an individual particle, though scattering experiments have given a strong indication that the theory is correct.

The weak interaction

The weak interaction has a very short range. This suggests that its gauge bosons are relatively massive, since a large mass, i.e. a high energy, would mean a short lifetime and therefore the exchange particles could only travel a small distance. The weak interaction has three gauge bosons, known as the intermediate vector bosons W^+, W^- and Z. These bosons were discovered in 1983 at CERN in Geneva.

The weak interaction acts on leptons and on hadrons. In fact it is the only force, other than gravity, which acts on neutrinos. This explains the fact that neutrinos are so reluctant to interact with anything.

Radioactive beta decay is due to the weak interaction (see Fig 18).

beta-minus decay: a neutron decays into a proton, emitting an electron and an antineutrino. The decay occurs via the weak interaction and is mediated by a W^- boson:

$n \rightarrow p + e^- + \bar{v}_e$

beta-plus (positron) decay: a proton decays into a neutron, emitting an electron neutrino and a positron. The decay occurs via the weak interaction and is mediated by a W^+ boson:

$p \rightarrow n + v_e + e^+$

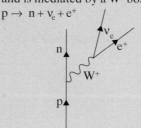

Fig 18
Examples of Feynman diagrams. All these reactions have been observed. They are all due to the weak interaction and are mediated by a W boson. Charge is conserved in all cases

electron capture: an atomic electron can be absorbed by a proton in the nucleus in a process called electron capture. The decay occurs via the weak interaction and is mediated by a W^+ boson:

$p + e^- \rightarrow n + v_e$

electron–proton collisions: an electron can collide with a proton, emitting a neutron and an electron neutrino. The reaction occurs via the weak interaction and is mediated by a W^- boson:

$p + e^- \rightarrow n + v_e$

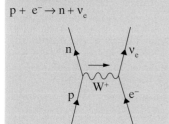

Examiners' Notes

At each junction (or node) in a Feynman diagram, charge is conserved. When you construct a Feynman diagram remember that, at each junction, the sum of the charges going into the point must equal the sum of the charges going out. The gauge boson should be shown either sloping up the page, or horizontal with an arrow to show direction.

Gravity

The gauge boson which carries the gravitational force is named the **graviton**. It is predicted to have zero rest mass and zero charge. It has never been detected.

Classification of particles

Leptons

Leptons are fundamental particles; they have no internal structure and are not affected by the strong interaction. There are 12 different particles in the lepton family. The most familiar lepton is the electron. Two other particles, the **muon** and the **tau**, are similar to the electron but more massive. Each has an associated neutrino (see Table 5). All these particles have an antiparticle of opposite charge.

The muon, μ^-, was discovered in cosmic-ray studies in 1937. It carries the same charge as the electron but its mass is about 207 times greater.

In 1962 it was shown that the neutrinos that accompany muons are not the same as electron neutrinos. The muon neutrino ν_μ and its antiparticle $\bar{\nu}_\mu$ are also fundamental particles which carry no charge.

In 1978 another member of the lepton family was discovered. The tau minus particle, τ^-, was observed by a team working on electron–positron collisions at Stanford in the USA. The tau particle has the same charge as the electron and the muon but is around 3500 times the mass of the electron. This new, heavier lepton has its own type of neutrino and antineutrino, the ν_τ and the $\bar{\nu}_\tau$.

Table 5
Leptons: each of these leptons has an antiparticle with the same mass, but opposite charge

Lepton	Symbol	Charge (in terms of proton charge)	Mass (in terms of electron mass)
electron	e^-	−1	1
electron neutrino	ν_e	0	near zero
muon	μ^-	−1	207
muon neutrino	ν_μ	0	near zero
tau	τ^-	−1	3500
tau neutrino	ν_τ	0	near zero

Leptons: a summary

- The **leptons**, and their antiparticles, **antileptons**, are believed to be fundamental particles.

- There are three negatively charged leptons: the **electron**, the **muon** and the **tau particle**.

- Each of these charged leptons has an associated **neutrino**.

- Leptons are not affected by the strong interaction.

Hadrons

In addition to the leptons, the particles now known as pi mesons, kaons and delta mesons had all been discovered by the 1960s. All of these have masses much larger than the leptons. This group of particles were referred to as **hadrons**.

Essential Notes

The term 'lepton' comes from the Greek word *leptos*, meaning 'small' or 'dainty'. 'Hadron' is from the Greek word *hadros*, meaning 'bulky' or 'thick'.

By the late 1960s a large number of hadrons had been discovered. Some carried charge and others did not. All except the proton were found to be unstable. Like radioactive atoms, after a certain time they decay into something else. In fact, all hadrons eventually decay into a proton. Even the neutron is unstable when it is free of the nucleus. The neutron decays with a half-life of around 11 minutes by emitting a beta particle:

$$n \rightarrow p + e^- + \bar{\nu}_e$$

Examiners' Notes

You need to know this interaction.

The hadrons are themselves divided into two groups, the **baryons** and the **mesons**. The baryons, which were orginally thought to be the heavier group, include the proton and the neutron and their antiparticles. The mesons include a large number of particles originally found in cosmic rays, which are now commonly created in collisions inside particle accelerators. The pi meson, or pion, was one of the first to be discovered. It exists in three different forms, positively charged, negatively charged and uncharged: π^+, π^- and π^0. Many other mesons have since been discovered, all of them unstable, usually with very short lifetimes. One of the most puzzling at first was the K meson, or kaon, which had a much longer lifetime than other, apparently similar mesons.

Conservation laws

A large number of hadron reactions have been studied. It became apparent that some reactions which appeared to be possible never took place. It seemed as if some reactions were forbidden. We now know that there are physical quantities which cannot change in particle reactions. These *conserved* quantities govern which reactions can take place.

Conservation of charge

Charge is a familiar idea. We know that charged bodies can exert a force on each other and that there are two 'types' of charge which we call positive and negative. Charge is extremely important in particle physics. Many hadrons, and leptons, carry a charge. It is usual to define the charge of the proton to be $+1$. Then the charge of the electron is -1, that of the positron $+1$ and that of the neutron 0. One of the rules that governs the interactions between particles is the conservation of charge (Table 6). No reactions that contravene this rule have ever been observed.

Definition

*The **conservation of charge** means that the total charge after a reaction is the same as the total charge before the reaction.*

	Before		After	
reaction	p + p	→	p + p + π^- + π^+	
charge Q	1 + 1 = 2		1 + 1 + (−1) + 1 = 2	allowed
reaction	p + π^-	→	p + π^+	
charge Q	1 + (−1) = 0		1 + 1 = 2	not allowed

Table 6
Particle interactions and conservation of charge Q

Essential Notes

'Baryon' is from the Greek word *barus*, meaning 'heavy'. This is because baryons were thought to be heavier than mesons. This is not always true.

Examiners' Notes

You need to learn the baryon numbers for the hadrons in Table 7.

Conservation of baryon number

There are some reactions allowed by charge conservation which have never been observed. This is because there are other conservation laws that place restrictions on which reactions can take place. One of these is the conservation of **baryon number**, B.

Hadrons can be classified by baryon number into **mesons**, **baryons** and **antibaryons** (Table 7). Mesons have a baryon number of 0, baryons 1 and antibaryons -1. Reactions between any of these hadrons can only occur if the baryon number is conserved. Just like charge, the total baryon number before the interaction has to be the same as the total baryon number afterwards (Table 8).

All other particles, i.e. the leptons and the gauge bosons, have a baryon number of 0.

Table 7
Baryon number B for some hadrons

Mesons $B = 0$	Baryons $B = 1$	Antibaryons $B = -1$
pi mesons (pions) π^-, π^-, π^0	proton p and neutron n	antiproton $\bar{p}$
K mesons (kaons) K^+, K^-, K^0	sigma particles Σ^-, Σ^-, Σ^0	antineutron $\bar{n}$

Table 8
Particle interactions and conservation of baryon number B

	Before		After	
reaction	p + p	→	p + p + n	
B	$1 + 1 = 2$		$1 + 1 + 1 = 3$	not allowed
reaction	p + p	→	$p + p + \pi^0$	
B	$1 + 1 = 2$		$1 + 1 + 0 = 2$	allowed

Conservation of strangeness

The rules of conservation of charge and conservation of baryon number do not fully explain why some reactions are never observed.

K mesons, or **kaons**, caused particular problems to particle physicists. Kaons appeared as the decay products of some neutral particles, but they always seemed to turn up in pairs. Kaons didn't appear individually, although charge and baryon number conservation would not prevent this. They also had an unusually long lifetime, 10^{-10} s, compared with other, apparently similar, hadrons, which had typical lifetimes of the order of 10^{-23} s.

There is another property that has to be conserved in hadron reactions. This property is called **strangeness**. All hadrons are given a strangeness number, S, of ±3, ±2, ±1 or 0. Strangeness has to be conserved in any reaction that takes place via the strong interaction (Tables 9 and 10).

Essential Notes

Unlike charge conservation and conservation of baryon number, there are some interactions where strangeness is not conserved. These are decays that take place via the weak interaction.

S = −2	S = −1	S = 0	S = +1
Ξ⁻ (xi minus)	Λ (lambda)	p (proton)	K⁺ and K⁰ (kaons)
Ξ⁰ (xi zero)	K⁻ (K minus) Σ^+, Σ^-, Σ^0 (sigma particles)	n (neutron) π^+, π^-, π^0 (pions)	

Table 9
Strangeness S for some hadrons

	Before		After	
reaction S	$p + \pi^-$ $0 + 0 = 0$	→	$K^0 + \Lambda^0$ $1 + (-1) = 0$	allowed
reaction S	$p + \pi^-$ $0 + 0 = 0$	→	$K^- + \Sigma^+$ $(-1) + (-1) = -2$	not allowed

Table 10
Particle interactions and conservation of strangeness S

Allowed reactions

All hadrons have fixed values for charge Q, baryon number B and strangeness S. A reaction between hadrons can only take place if the reaction conserves these numbers.

When strange particles, i.e. those with non-zero values of strangeness, are produced, they have to appear in pairs in order to conserve strangeness. When they decay, however, they do so by the weak interaction and strangeness is *not* conserved. In these reactions strangeness changes by ±1.

Example

Which of these reactions is not allowed?

(a) $\pi^+ + p \rightarrow K^+ + n$

(b) $K^+ + \bar{p} \rightarrow \pi^0$

(c) $p + p \rightarrow p + p + \pi^0$

Answer

(a) Cannot occur, as it violates charge conservation and strangeness conservation.

(b) Cannot occur, as it violates baryon number and strangeness conservation.

(c) Can occur.

Examiners' Notes

Check that Q, B and S are the same on both sides of the equation.

Quarks and antiquarks

In the 1960s scattering experiments carried out at Stanford in the USA revealed the structure inside hadrons.

The Stanford Linear Accelerator Center (SLAC) in California could accelerate electrons to an energy of around 6 GeV, high enough to probe the structure of nucleons, i.e. to look inside protons and neutrons. The SLAC experiments found that a significant proportion of high-energy electrons were scattered through a large angle. This indicated that the neutrons and protons are not particles of uniform density, but have pointlike charged particles within them (Fig 19).

Fig 19
Electron scattering by quarks within a baryon

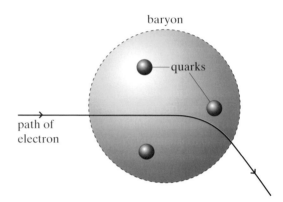

Essential Notes

The name 'quark' was given to these particles by Murray Gell-Mann. He took the term from 'Three quarks for Muster Mark', a quotation from **Finnegans Wake** by James Joyce.

The SLAC results confirmed a theory put forward by Murray Gell-Mann and George Zweig a few years earlier. Gell-Mann had grouped the hadrons together in families (Fig 20). The patterns could be explained by supposing that all hadrons were composed of smaller constituents, named **quarks**. The SLAC experiments confirmed that hadrons are not fundamental particles but are composed of combinations of different types of quark.

Fig 20
A family of baryons. Gell-Mann called the pattern the 'eightfold way', a term from Buddhism

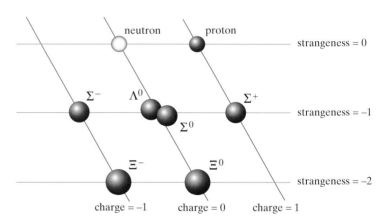

The eight baryons in Fig 20 all have a baryon number of 1. There is a similar grid for their antiparticles (baryon number -1). The Λ^0 and Σ^0 differ only in their energy.

Gell-Mann suggested that there were three different quarks, labelled 'up' (u), 'down' (d) and 'strange' (s). Each of these quarks has a particular mass and values for charge, baryon number and strangeness. Each quark has a corresponding **antiquark** of exactly equal mass but opposite values for charge, baryon number and strangeness (Table 11).

Table 11
Properties of quarks and antiquarks

Quark	Baryon number B	Charge Q	Strangeness S	Antiquark	Baryon number B	Charge Q	Strangeness S	Mass (GeV/c^2)
up, u	$\frac{1}{3}$	$\frac{2}{3}$	0	$\bar{u}$	$-\frac{1}{3}$	$-\frac{2}{3}$	0	0.005
down. d	$\frac{1}{3}$	$-\frac{1}{3}$	0	$\bar{d}$	$-\frac{1}{3}$	$-\frac{1}{3}$	0	0.01
strange, s	$\frac{1}{3}$	$-\frac{1}{3}$	-1	$\bar{s}$	$-\frac{1}{3}$	$-\frac{1}{3}$	$+1$	0.2

Using the quark model it is possible to describe all hadrons in terms of combinations of quarks and antiquarks (Fig 21). **Baryons** are combinations of three quarks; **antibaryons** are combinations of three antiquarks. **Mesons** are composed of a quark and an antiquark, not necessarily the same type.

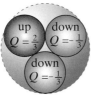

proton, $Q = 1$ K$^+$meson, $Q = 1$ neutron, $Q = 0$ π^- meson, $Q = -1$

Fig 21
Proton, K$^-$ meson, neutron and π^- meson, showing quark structure

The properties of each hadron can be explained in terms of the quarks that it is made from. The total charge on the hadron is the sum of the quark charges. The same can be said for the total baryon number and strangeness. Tables 12 to 17 summarise the quark structures of some hadrons.

Examiners' Notes

You need to know the structures in Tables 12 to 17.

	Up, u	Up, u	Down, d	Proton, p
charge Q	$\frac{2}{3}$	$\frac{2}{3}$	$-\frac{1}{3}$	1
baryon number B	$\frac{1}{3}$	$\frac{1}{3}$	$\frac{1}{3}$	1
strangeness S	0	0	0	0

Table 12
Quark structure of a proton:
p = u + u + d

	Antiup, ū	Antiup, ū	Antidown, d̄	Antiproton, p̄
charge Q	$-\frac{2}{3}$	$-\frac{2}{3}$	$\frac{1}{3}$	-1
baryon number B	$-\frac{1}{3}$	$-\frac{1}{3}$	$-\frac{1}{3}$	-1
strangeness S	0	0	0	0

Table 13
Quark structure of an antiproton:
p̄ = ū + ū + d̄

	Up, u	Down, d	Down, d	Neutron, n
charge Q	$\frac{2}{3}$	$-\frac{1}{3}$	$-\frac{1}{3}$	0
baryon number B	$\frac{1}{3}$	$\frac{1}{3}$	$\frac{1}{3}$	1
strangeness S	0	0	0	0

Table 14
Quark structure of a neutron:
n = u + d + d

	Antiup, ū	Antidown, d̄	Antidown, d̄	Antineutron, n̄
charge, Q	$-\frac{2}{3}$	$\frac{1}{3}$	$\frac{1}{3}$	0
baryon number, B	$-\frac{1}{3}$	$-\frac{1}{3}$	$-\frac{1}{3}$	-1
strangeness, S	0	0	0	0

Table 15
Quark structure of an antineutron:
n̄ = ū + d̄ + d̄

Table 16

Quark structure of a π^- meson:

$\pi^- = \bar{u} + d$

	Antiup, $\bar{u}$	Down, d	π^-
charge Q	$-\frac{2}{3}$	$-\frac{1}{3}$	-1
baryon number B	$-\frac{1}{3}$	$\frac{1}{3}$	0
strangeness S	0	0	0

Table 17

Quark structure of a K^+ meson:

$K^+ = u + \bar{s}$

	Up, u	Antistrange, $\bar{s}$	K^+
charge Q	$\frac{2}{3}$	$\frac{1}{3}$	1
baryon number B	$\frac{1}{3}$	$-\frac{1}{3}$	0
strangeness S	0	1	1

Table 18

Description of pions and kaons in the simple quark model

π^+	$u\bar{d}$
π^-	$\bar{u}d$
π^0	$u\bar{u}$ or $d\bar{d}$
K^+	$u\bar{s}$
K^-	$\bar{u}s$
K^0	$d\bar{s}$
$\bar{K}^0$	$\bar{d}s$

Only two types of quark, the up and the down, are needed to account for the properties of the neutron and proton which together make up almost all of everyday, observable matter. The two antiquarks $\bar{u}$ and $\bar{d}$ are needed to explain the existence of the antiproton and the antineutron.

The properties of a meson are the sum of the properties of its constituent quarks. A meson's antiparticle is the opposite combination of quark and antiquark. For example the pi plus meson is formed from the two quarks $u\bar{d}$, whilst its antiparticle, the pi minus, is formed from the opposite combination $\bar{u}d$ (Table 18). The K^- meson is formed from the quarks $\bar{u}s$.

The quark model has been very successful in describing and predicting the properties of hadrons. In beta-minus emission, for example, a neutron decays to a proton, emitting an electron and an antineutrino in the process. The quark theory tells us that inside the neutron a down quark has changed into an up quark, emitting an electron and an antineutrino (Fig 22).

Positron emission occurs when a proton inside a nucleus changes into a neutron. One of the up quarks has changed into a down quark. A positron and an electron neutrino are emitted.

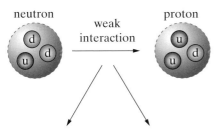

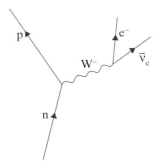

beta particle e⁻   antineutrino $\bar{\nu}_e$

or, as a Feynman diagram,

Fig 22
Beta-minus emission

or, in terms of quarks,

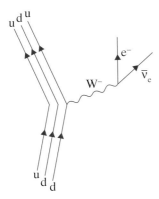

3.1.2 Electromagnetic radiation and quantum phenomena

Essential Notes

A **transverse wave** has oscillations at right angles to its direction of travel, whereas a **longitudinal wave** (like sound) has oscillations that are parallel to the direction of travel.

Electromagnetic waves

Electromagnetic waves are emitted by the oscillation of charged particles, such as an electron. The oscillation sets up varying electric and magnetic fields which travel through space. The electric and magnetic fields are at right angles to each other and to the direction of travel of the waves (Fig 23). The wave propagates through space as a transverse wave, without the need for any supporting medium.

Fig 23
An electromagnetic wave

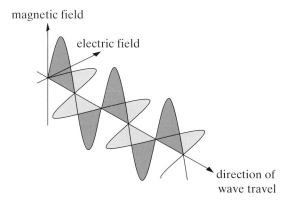

All electromagnetic waves travel at the same speed in a vacuum, that is, 2.98×10^8 m s^{-1}. However, the properties of the wave and the way that it interacts with matter depend on its wavelength.

Definition

*The distance between any two identical points on a wave is the **wavelength**, λ.*

This is measured in metres. For a pure sine wave it is the distance between any two adjacent crests or troughs.

Definitions

*The time taken for a wave to go through one complete oscillation is the **period**, T.*

*The number of oscillations per second, f, is the **frequency**.*

The frequency is measured in hertz (Hz). Frequency $= \dfrac{1}{\text{period}}$, that is,

$$f = \frac{1}{T}$$

Since speed = distance divided by time, the speed that the wave travels at, c, is given by $c = \dfrac{\lambda}{T}$ or $c = \lambda \times \dfrac{1}{T}$.

So
$$c = f\lambda$$

Light

Light is an electromagnetic wave. Visible light has a wavelength range from about 400 nm (violet) to around 700 nm (red).

Example

Find the frequency range of visible light.

Answer

Since $c = f \times \lambda$,

$$f = \frac{c}{\lambda}$$

If $\lambda = 400 \times 10^{-9}$ m, then

$$f = \frac{3 \times 10^8}{400 \times 10^{-9}} = 7.5 \times 10^{14} \text{ Hz}$$

If $\lambda = 700 \times 10^{-9}$ m, then

$$f = \frac{3 \times 10^8}{700 \times 10^{-9}} = 4.3 \times 10^{14} \text{ Hz}$$

Essential Notes

1 **nanometre** (1 **nm**) = 1×10^{-9} m

Different wavelengths, or frequencies, of light give us the impression of different colours. Larger-amplitude waves increase the intensity (brightness) of the light. Like the rest of the electromagnetic spectrum, light can be reflected, refracted and diffracted. These wave phenomena are covered in detail in Unit 2 of *AS Physics*.

The photoelectric effect

The **photoelectric effect** was discovered towards the end of the 19th century. Experiments showed that electrons could be emitted from the surface of a metal by illuminating the metal with light. However, some of the experimental results seemed completely at odds with the wave theory of light.

- The electrons are only emitted from the surface of the metal if the light is above a certain frequency. For example, if zinc is illuminated with visible light, no electrons are emitted. It is only when ultraviolet light is used that there is any effect. Every metal has its own particular light frequency, known as the **threshold frequency**, below which there is no photoemission.

- The electrons are emitted with a range of different kinetic energies from zero up to a maximum value. The maximum kinetic energy depends on the frequency of the light, not on the intensity. A faint ultraviolet glow would cause the emission of more energetic electrons than an intense red laser beam.

- If the light is above the threshold frequency, then the number of electrons emitted per second is proportional to the intensity of the light.

- If the light is above the threshold frequency, photoemission starts immediately the light falls on the surface, no matter how low its intensity.

Electrons are held by electrostatic forces onto the surface of the metal. The light has to provide enough energy to rip an electron free from the metal surface.

> **Definition**
>
> *The energy needed to remove an electron from the surface of a metal is called the **work function**, denoted by ϕ. This is usually given in electron volts, eV, and depends on the type of metal.*

The wave theory of light says that if a light wave hasn't got enough energy to release an electron, then you need a higher-amplitude wave, i.e. brighter light. But this doesn't work. If the light wave is below the threshold frequency, it doesn't matter how intense it is, there will be no photoemission.

Einstein explained the photoelectric effect by using the idea of **photons**. He realised that light is absorbed in discrete packets or **quanta** of electromagnetic energy, called photons. When a photon strikes a metal surface, it is absorbed either totally or not at all. So when a photon strikes the surface of a metal and collides with an electron, it will only dislodge an electron if its energy, E, is larger than the work function.

Photoemission only occurs when

$$E > \phi$$

or, since

$$E = hf$$

where h is the Planck constant (see p. 16), only when

$$hf > \phi$$

When photoemission *just* occurs, the threshold frequency f_0 is given by

$$f_0 = \frac{\phi}{h}$$

Above the threshold frequency, the photon carries more than enough energy to release an electron. The excess energy goes into the kinetic energy of the emitted electron. Einstein's photoelectric equation expresses this in terms of the conservation of energy:

$$\text{energy of incident photon} = \text{energy needed to remove the electron (work function)} + \text{kinetic energy of the emitted electron}$$

$$hf = \phi + E_k$$

E_k is the maximum kinetic energy of the electron; if the electron has been emitted from deeper within the metal surface it may have a lower value of kinetic energy.

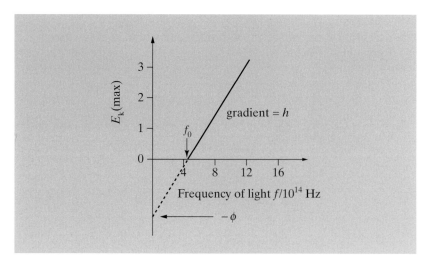

Fig 24
Maximum kinetic energy of emitted electron versus frequency of incident light:
$E_k(\max) = hf - \phi$

Essential Notes

This graph shows exactly what Einstein predicted: that the maximum energy of the released electrons is proportional to the frequency of the light. The constant of proportionality (the gradient of the graph) is equal to h.

Example

Light of wavelength 330 nm is incident on a metal which has a threshold frequency of 5.0×10^{14} Hz. Find the maximum kinetic energy of the emitted electrons.

Answer

The energy of the incident photons is
$$E = hf = \frac{hc}{\lambda} = 6.03 \times 10^{-19}\,\text{J}$$

The work function of the metal is
$$hf_0 = 6.6 \times 10^{-34} \times 5.0 \times 10^{14} = 3.32 \times 10^{-19}\,\text{J}$$

Using Einstein's photoelectric equation
$$hf = \phi + E_k$$
$$E_k = hf - \phi = 6.03 \times 10^{-19} - 3.32 \times 10^{-19} = 2.7 \times 10^{-19}\,\text{J or } 1.7\,\text{eV}$$

Collisions of electrons with atoms

The electron volt
In atomic and nuclear physics physicists use the **electron volt** as a convenient unit of energy.

> **Definition**
>
> An **electron volt** (eV) is the amount of energy gained by an electron as it accelerates through a potential difference of 1 volt. $1\,\text{eV} = 1.6 \times 10^{-19}\,\text{J}$

Cathode ray tubes such as the one used by Thomson (see p. 4) achieved electron energies of around 1 keV. The Large Hadron Collider at CERN, in Geneva, can accelerate protons to energies of several TeV ($1\,\text{TeV} = 10^{12}\,\text{eV}$).

Line spectra

When an electric current is passed through a vapour of an element, the electrons collide with atoms of the vapour and light is given off. If the light is observed through a diffraction grating, each element is found to have its own set of bright emission lines. This is called a **line spectrum**. Fig 25 shows the lines for hydrogen. For any specific element the spectral lines are always at the same frequency. To explain this we need to look again at our model of the atom.

Fig 25
Line spectrum for hydrogen. Each element has its own characteristic line spectrum

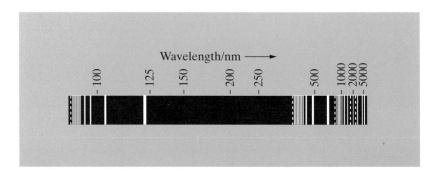

Rutherford's model of the hydrogen atom has one electron orbiting a very small, dense, positively charged nucleus (see p. 8). There is a problem with this model. All charged particles emit radiation when they accelerate. The orbiting electron is accelerating towards the centre of its orbit as it constantly changes direction. According to the laws of classical physics the electron should be radiating energy all the time. As it radiates it should lose energy, eventually spiralling down towards the nucleus. This is rather like an artificial satellite that has dropped into a low orbit around the Earth. As the satellite passes through the upper atmosphere it loses energy and so it will inevitably drop further and spiral down towards the Earth's surface.

Niels Bohr suggested that the electron could travel in certain allowed orbits (Fig 26) without losing energy. He called these allowed orbits 'stationary states'. When the electron is in an allowed orbit it does not radiate, but stays at a constant energy.

Essential Notes

The energy levels are negative because the electron is in a bound state – it is tied to the atom. The energy value of each allowed orbit tells you how much energy is needed to free the electron from the atom. By this convention higher energy levels are less negative; an electron with zero energy is just free of the atom.

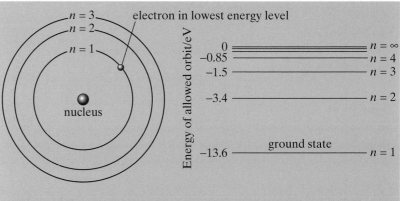

The ground state, $n = 1$, is the lowest energy level. An electron in this level needs an energy transfer of 13.6 eV to free it from the atom.

Fig 26
Allowed orbits and energy levels in Bohr's hydrogen atom

Energy levels and photon emission

Bohr proposed that an electron in an atom can only emit or absorb energy as it moves from one allowed orbit to another (Fig 27). This idea helped to explain the existence of line spectra. Light is emitted from atoms when electrons lose energy, but in Bohr's atom electrons can only lose energy in specified amounts as they jump down the energy levels. An electron can only move from one allowed state to another by gaining or losing exactly the right amount of energy. This is called an **electron transition**. That is why only certain frequencies of light appear in the line spectrum.

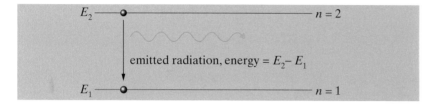

Fig 27
Electron transition

Each time an electron falls to a lower energy level it loses energy. This energy is radiated as a photon of frequency f. So the energy of the emitted photon is

$$E_2 - E_1 = \Delta E = hf$$

Bohr's model of the atom was successful in explaining the existence of line spectra. Although the energy level diagrams are more complicated for atoms other than hydrogen, the same principles apply.

Ionisation and excitation

When the electrons are in their lowest-energy orbits, an atom is said to be in its **ground state**. The lowest allowed orbit for a hydrogen atom has an energy of -13.6 eV. When a hydrogen atom is in the ground state, its electron cannot lose any more energy. The ground state is the preferred state for an atom, but electrons can move to higher energy levels if the atom absorbs the correct amount of energy. This process could be caused by the absorption of a photon of radiation of the right wavelength or by a collision with another electron (Fig 28).

Definitions

Excitation is when an atomic electron moves to a higher energy level.

Ionisation is when an electron gains so much energy that its total energy becomes positive. This means that it becomes free of the atom.

Example

(a) A fast-moving electron with a kinetic energy of 15.0 eV collides with an electron in a hydrogen atom. Explain what is likely to happen.

(b) What would happen if the incident electron had an energy of 10.5 eV?

Fig 28
Excitation by (a) absorption of a photon and (b) collision of an electron; (c) shows ionisation

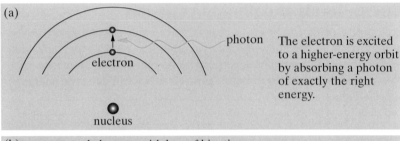

(a)

photon — The electron is excited to a higher-energy orbit by absorbing a photon of exactly the right energy.

electron

nucleus

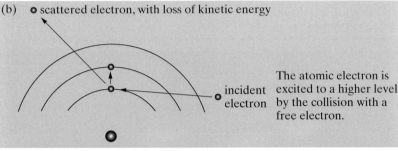

(b) scattered electron, with loss of kinetic energy

incident electron — The atomic electron is excited to a higher level by the collision with a free electron.

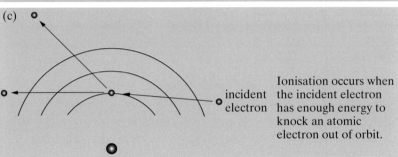

(c)

incident electron — Ionisation occurs when the incident electron has enough energy to knock an atomic electron out of orbit.

(c) What would happen if the incident electron had an energy of 8 eV? (You will need to refer to the energy level diagram for the hydrogen atom in fig 26.)

Answer

(a) Assuming the hydrogen atom is in its ground state, its electron needs only 13.6 eV to free it from the atom. The incident electron will transfer some of its kinetic energy to the atomic electron. The collision will cause ionisation and the ejected electron will gain kinetic energy.

(b) The incident electron has enough energy to excite the electron in the hydrogen atom to a higher energy level. The atomic electron would be excited to the $n = 2$ orbit, which requires $13.6 - 3.4 = 10.2$ eV. There would be $10.5 - 10.2 = 0.3$ eV of energy left as kinetic energy of the incident electron. This is called **inelastic scattering** because the incident electron has lost kinetic energy.

(c) The incident electron does not have enough energy to raise the atomic electron to a higher energy level. The atomic electron cannot gain any energy from the incident electron, which scatters off the atom without losing any energy. This is known as **elastic scattering**.

The fluorescent lamp

When an electric current is passed through a fluorescent lamp, electrons collide with atoms of mercury vapour. If an electron has sufficient energy, greater than 6.7 eV, the collision will excite an electron in the mercury atom to a higher energy level. As the atomic electron returns to its original state it emits a photon of ultraviolet light. The photons of ultraviolet light are absorbed by atoms in the phosphor coating on the inside of the glass lamp, and electrons in these atoms are excited to higher energy states. As these electrons fall to lower energy states they emit photons of visible light. The phosphors are carefully chosen to have the right energy levels to produce the required colour of light from the lamp.

Wave–particle duality

Photons or waves?

The wave theory successfully explains the way that light is reflected and refracted and is important in interpreting the phenomena of diffraction and interference. However, the wave theory cannot describe the photoelectric effect, nor black-body radiation curves (p. 15). These have been explained by thinking of light as a stream of massless particles called photons, which carry energy. This is an example of what is referred to as **wave–particle duality**. There is no real contradiction here. The wave and particle theories are complementary to each other. The wave theory gives us an excellent way of picturing what happens as light passes from one place to another, whilst the particle theory is useful in describing how light interacts with electrons (Fig 29).

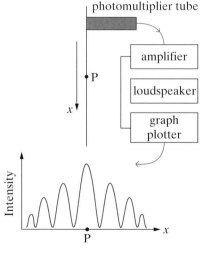

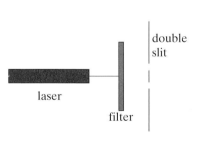

A photomultiplier tube uses the photoelectric effect to detect light photons. It therefore relies on the particle nature of light, yet it detects an interference pattern which can only be caused by waves. If the filter is dense enough, the loudspeaker lets you hear the photons arriving, one by one.

Fig 29
Wave–particle duality experiment

Particles or waves?

It isn't only light that shows aspects of wave and particle behaviour. Electrons, which we have so far treated as point particles, can be made to diffract. Louis de Broglie suggested, in 1924, that 'all material particles should have a wave nature'. He predicted that a particle of momentum p should have a wavelength of λ given by

$$\lambda = \frac{h}{p} \quad \text{or} \quad \lambda = \frac{h}{mv}$$

where h is the Planck constant. This is called the **de Broglie wavelength**.

This idea can be tested by trying to diffract electrons through a suitable aperture. Diffraction effects become noticeable when the size of the aperture is of the same magnitude as the wavelength of the waves. An electron accelerated across an electric potential of 5 kV will reach a velocity of around 4.2×10^7 m s^{-1}. This gives it a momentum of 3.8×10^{-23} kg m s^{-1}. According to de Broglie's equation, the electron has a wavelength of

$$\lambda = \frac{h}{p} = \frac{6.6 \times 10^{-34}}{3.8 \times 10^{-23}} = 1.7 \times 10^{-11}\,\text{m}$$

This is about the size of the gaps between layers of atoms.

In 1928, four years after de Broglie put forward his theory, George Thomson (the son of J.J. Thomson) produced an electron diffraction pattern by firing high-speed electrons at a gold foil. The emerging electron beam showed the same variation in intensity as light that had passed through a diffraction grating.

Today electron microscopes, which rely on the wave nature of electrons, are in common use. There are also microscopes that use protons and even ions. Since these particles are more massive and carry more momentum, their de Broglie wavelength is even smaller, which gives improved resolution.

Essential Notes

The electron microscope helps us to see much finer detail than is possible using a light microscope. This is because it is diffraction that limits our ability to see fine detail. Electrons diffract very little because they have a very short wavelength.

3.1.3 Current electricity

Charge, current and potential difference

Definition

*An electric **current** is defined as the rate at which electrically charged particles pass through a point in a circuit.*

The size of the current is measured in coulombs per second or amperes (amps).

$$1 \text{ coulomb per second} = 1 \text{ ampere}$$

In metallic **conductors** the charge-carriers are electrons, which move from the negative terminal of the d.c. supply towards the positive terminal. Confusion can arise, because current is normally shown as moving from the positive terminal towards the negative terminal. This is referred to as '**conventional current**'.

Essential Notes

All current arrows on wires and component symbols point in the conventional current direction, i.e. in the direction that a positive charge would move.

Definition

The size of the current is defined mathematically by

$$I = \frac{\Delta Q}{\Delta t}$$

where I = current in amps (A), Q = charge in coulombs (C), t = time in seconds (s). ΔQ = change in charge and Δt = change in time.

To make current flow, a **potential difference** (p.d.) must exist.

Definition

*A **potential difference** is defined as the electrical energy transferred or converted per unit of charge passing between the two points.*

Potential difference is measured in joules per coulomb or volts.

1 joule per coulomb = 1 volt

Definition

The size of the potential difference (p.d.) is defined mathematically by

$$V = \frac{W}{Q}$$

where V = p.d. in volts, W = work (energy) in joules, Q = charge in coulombs.

Essential Notes

$V = \dfrac{W}{Q}$ can be rearranged to

give $Q = \dfrac{W}{V}$ and $W = VQ$

A charge gains energy when it passes through a cell. It 'releases' the gained energy as it passes through components in a circuit (e.g. a lamp or resistor); i.e. a p.d. exists across the component. Thus, both a cell and a component have a p.d. across them when charge flows in a circuit.

Charges face opposition when they flow around a circuit. This is called **resistance** and it is measured in ohms (Ω). The potential difference needed to make a current flow in a circuit depends on the resistance in the circuit. The bigger the resistance, the more p.d. is required to make a certain current flow.

Definition

Resistance is defined by the equation

$$R = \frac{V}{I}$$

where I = current in amps, V = p.d. in volts, R = resistance in ohms.

This expression can be rearranged to give:

$$V = IR \quad \text{or} \quad I = \frac{V}{R}$$

Table 19 shows some common multiples and submultiples of electrical units.

Table 19
Electrical units

Prefix	Example	Symbol	Quantity
milli-	milliamps	mA	10^{-3} amps
micro-	microamps	µA	10^{-6} amps
milli-	millivolt	mV	10^{-3} volts
kilo-	kilohm	kΩ	10^{3} ohms
mega-	megohm	MΩ	10^{6} ohms

Examiners' Notes

It is important to use the correct units. Table 19 will help you to convert other units to volts, amps and ohms. A common error is to confuse milliamps and microamps.

Example

In a conductor the charge carriers each have a charge of 1.6×10^{-19} C.

(a) Calculate the number of charge carriers passing a point in the conductor per second if the current is 4.0 µA.

(b) Calculate the p.d. generated by the charges across a 1000 ohm resistor.

(c) Calculate the work done per charge carrier.

Answer

(a) $Q = It = 4.0 \times 10^{-6} \times 1$
$ = 4.0 \times 10^{-6}$ C

$$\text{Number of charge carriers} = \frac{\text{total charge}}{\text{charge on charge carrier}}$$

$$= \frac{4.0 \times 10^{-6}\ \text{C}}{1.6 \times 10^{-19}\ \text{C}}$$

$$= 2.5 \times 10^{13} \quad \text{(This value has no units.)}$$

(b) $V = IR = 4.0 \times 10^{-6} \times 1000$
$ = 4.0 \times 10^{-3}$ V (or 4 mV)

(c) $W = VQ = 4.0 \times 10^{-3} \times 1.6 \times 10^{-19}$
$ = 6.4 \times 10^{-22}$ J

Current/voltage characteristics

The circuit in Fig 30 can be used to investigate how the potential difference across a component affects the current through it. The values of current and p.d. can be plotted on a graph which is known as a current/voltage characteristic.

Fig 30
Circuit used to produce current/voltage characteristics of components

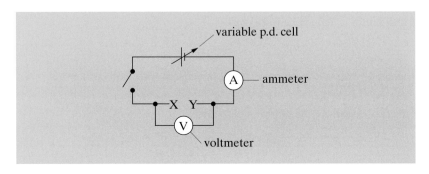

- The component under test is placed between points X and Y so that the circuit is complete when the switch is closed.

- By varying the supply p.d. a range of current and p.d. values can be recorded for the component using the ammeter and voltmeter readings.

- The battery is reversed and the supply p.d. varied over the same range to produce a second set of current and p.d. values.

- A graph of the results can then be drawn, which is the 'characteristic curve' for the component.

Resistor or wire (ohmic conductor)

When the current is plotted against the p.d., a straight line graph is obtained (Fig 31).

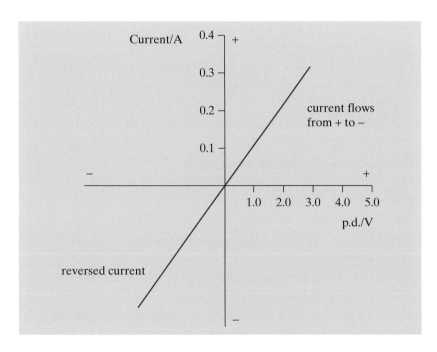

Fig 31
Current plotted against p.d. for an ohmic conductor

Examiners' Notes

When drawing sketch graphs, try to label axes with an appropriate scale and units. In this case we are just showing general characteristics.

The current and p.d. are directly proportional to each other (straight line through the origin) when the current flows in either direction. The conductor is said to follow **Ohm's Law**. We will deal with Ohm's Law in greater detail on p. 41.

Semiconductor diode

In the case of a semiconductor **diode**, the shape of the curve obtained depends on the direction in which the current is flowing (Fig 32).

When the diode is **forward biased** (arrow facing the direction of conventional current):

- between 0 V and about 0.7 V, the diode offers a large resistance to current

- between about 0.7 V and 1 V the resistance of the diode falls rapidly and a large current flows – this is shown by the steep rise in the graph.

Essential Notes

The diode is a one-way device. It acts like a valve.

Fig 32
Current plotted against p.d. for a
semiconductor diode

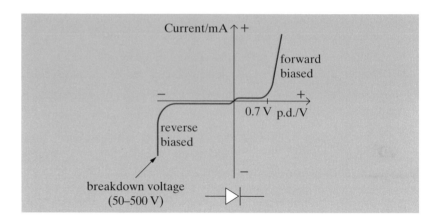

When the diode is **reverse biased** (arrow facing the opposite direction to conventional current) and the voltage is steadily increased:

- the diode offers high resistance, so very little or no current flows

- at the breakdown voltage, typically between 50 and 500 V, a large current flows

- most diodes cannot recover and are destroyed by the heating effect of the large current.

Filament lamp (non-ohmic conductor)

When a filament lamp is connected between X and Y in Fig 30, and the voltage is steadily increased:

- the graph becomes less steep

- the p.d. and current do not increase proportionally because the current heats the filament

- an increase in the temperature of the wire increases the resistance of the filament and so decreases the rate of increase of current with p.d.

The curve is symmetrical on either side of the origin, showing that the lamp behaves in the same way for current flowing through it in either direction (Fig 33).

Fig 33
Current plotted against p.d. for a
filament lamp

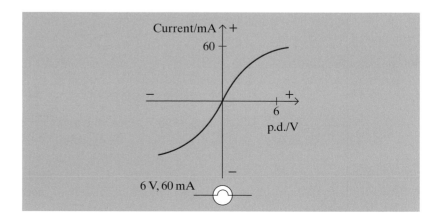

Producing current/voltage characteristics using a data logger

The same characteristics can be produced automatically, using a voltage sensor (V) and current sensor (A). These, together with a data logger (D), capture data which is then fed into the computer for analysis. A typical set-up is shown in Fig 34.

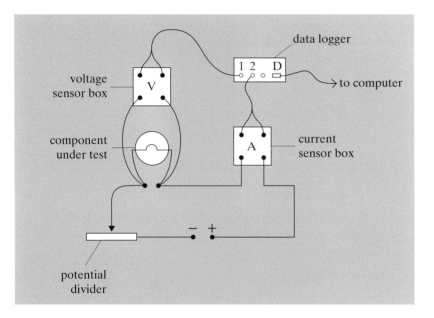

Fig 34
Data capture with a data logger

The potential difference is varied across the component under investigation (wire, resistor, lamp or diode) using a potential divider (see p. 51), and the current is recorded. The data logger software is then used to display the collected data in tabular and graphical form.

Ohm's Law

Ohm's Law is a special case and only applies to certain components in certain conditions.

> **Definition**
>
> *Ohm's Law* states that the current in a conductor is directly proportional to the p.d. across it,
>
> $$I \propto V$$
>
> *provided that the temperature and other physical conditions remain the same.*

Essential Notes

The equation
$$R = \frac{V}{I}$$
defines resistance. This equation can always be used to calculate R for any conductor when a particular current flows.

The current/voltage characteristics on pp 39–40 can show clearly whether or not a component obeys Ohm's Law. The graphs in Figs 31, 32 and 33 show that the resistor/wire obeys Ohm's Law, while the semiconductor diode and filament lamp do not.

Examiners' Notes

Do not confuse cross-sectional area with diameter,

$$A = \pi \left(\frac{d}{2}\right)^2$$

Doubling the diameter will reduce the resistance by a factor of 4.

Resistivity

Two factors which affect the resistance of a conductor are its length and its cross-sectional area.

- Resistance $\propto$ length ($R \propto l$) so doubling length doubles the resistance.

- Resistance $\propto = \dfrac{1}{\text{area}}$ $\left(R \propto \dfrac{1}{A}\right)$

  so doubling the cross-sectional area halves the resistance.

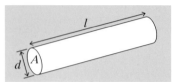

Fig 35

Examiners' Notes

The unit of resistivity is the ohm metre (Ω m). A common error in exams is to express it as Ωm^{-1}.

Definition

*The **resistivity** ρ of a material is given by:*

$$\rho = \frac{AR}{l}$$

This equation can be rearranged to give:

$$R = \frac{\rho l}{A}$$

where R is the resistance of the conductor, l is its length and A is its cross-sectional area. The resistivity ρ is a constant of the material from which the conductor is made and is measured in ohm metres (Ω m).

Examiners' Notes

Tolerance is usually applied when reading values from a graph, appropriate to the scale given.

Example

The graph shows the result of measuring how potential difference affects the current for a wire X with the cross-sectional area shown on the graph.

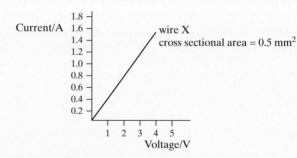

Fig 36

(a) Calculate the resistance of wire X.

(b) Calculate the resistivity of wire X if its length is 1.5 metres.

(c) Y is made from the same material as X. The resistance of 1.5 metres of Y is half that of X for the same length. Calculate the diameter of Y.

Answer

Examiners' Notes

Be careful with significant figures and quote answers to an appropriate value indicated by the data given in the question. A common error is to quote answers to one or two decimal places – this is not the same as significant figures.

(a) Measure the inverse gradient of the graph, e.g. $R = \dfrac{V}{I} = \dfrac{3}{1.1} = 2.73\ \Omega$

(b) Rearrange $R = \dfrac{\rho l}{A}$

$$\rho = \frac{RA}{l} = \frac{2.73 \times 0.5 \times 10^{-6}}{1.5} = 9.1 \times 10^{-7} \Omega\ m$$

(c) R for Y is $\dfrac{2.73}{2} = 1.37\,\Omega$

Rearrange $R = \dfrac{\rho l}{A}$

$A = \dfrac{\rho l}{R} = \dfrac{9.1 \times 10^{-7} \times 1.5}{1.37} = 9.96 \times 10^{-7}\ \text{m}^2$

$A = \pi\left(\dfrac{d}{2}\right)^2$

Rearranging, $d = 2\sqrt{\dfrac{A}{\pi}} = 1.1 \times 10^{-3}\ \text{m or } 1.1\ \text{mm}$

Examiners' Notes

Be careful when converting mm^2 to m^2; divide by 10^6 and not 10^3.

Measuring resistivity

The resistivity of a material in the shape of a wire can be measured using the apparatus in Fig 37.

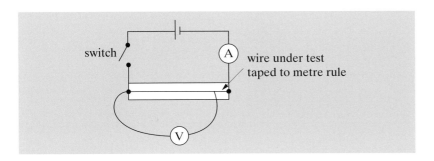

Fig 37
Measuring the resistivity of a material in the shape of a wire

- Start by measuring 100 cm of the wire under test. Tape the wire on to a metre rule, to avoid any kinks or twists. Connect the wire to the circuit using crocodile clips.

- Record, in a table, the p.d. displayed on the voltmeter and the current displayed on the ammeter, for this length of wire.

- Move the voltmeter connection along the wire in the range 100 cm to 30 cm and record the p.d. and current for each length.

- Calculate the resistance of wire for each recorded length using $R = \dfrac{V}{I}$.

- Measure the diameter of the wire, several times over its length, using a micrometer, to obtain a value for the mean diameter.

- Use the mean diameter to calculate the cross-sectional area using
  $A = \pi\left(\dfrac{d}{2}\right)^2$

- Plot resistance (y axis) against length (x axis) (Fig 38).
  $y = mx + c$ (equation of a straight line)

  $R = \left(\dfrac{\rho}{A} \times l\right) + 0$

- Use gradient = ρ/A to calculate ρ.

Essential Notes

Avoid large currents which will heat the wire and increase the resistance.

Essential Notes

A multimeter set on the ohms range could be used to measure resistance directly, instead of using a battery, ammeter and voltmeter. However, the ohms range usually has an uncertainty of $\pm 1\,\Omega$.

Fig 38
Graph showing resistance against
length for a wire

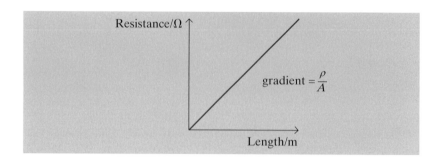

Temperature and resistance in conductors and thermistors

Temperature *always* affects conduction, no matter whether the material is a conductor, an insulator or a semiconductor. In conductors the resistance increases as the temperature increases (Fig 39).

Fig 39
Graph showing the increase in
resistance of a conductor with
increased temperature

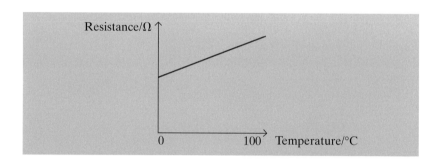

Metal wires and resistors have free electrons that move through the metal when a p.d. is applied, causing a current to flow. The metal also has vibrating positive ions. Electrons collide with these ions, causing the wire to have resistance to current.

As the temperature of the wire increases, the positive ions and electrons both absorb the heat energy, causing the ions to vibrate with greater amplitude and the electrons to move faster. Both of these effects result in a greater number of collisions between electrons and ions, i.e. the resistance of the conductor increases. However, the gradient of the graph is not very steep, showing that resistance does not change greatly with temperature.

In the case of **thermistors** the resistance decreases significantly as temperature increases (Fig 40).

Fig 40
In thermistors, resistance decreases
as temperature increases

Essential Notes

A thermistor is a device used
for temperature measurement
and control. The circuit symbol
for a thermistor is

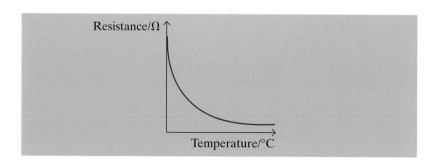

Small increases in temperature produce *large* changes in resistance of the thermistor. The thermistor is made from semiconductor material and therefore has few free electrons to produce a current. As the temperature of the thermistor increases, the thermal energy is enough to release further electrons from the ions to make the material conductive – resistance decreases.

Care is needed when passing current through thermistors. Currents produce heat and this decreases the resistance of the thermistor, allowing more current to flow. This further heats the thermistor, producing further resistance changes and the process can continue until the component overheats and burns out or melts.

Superconductors

If the temperature of a conductor is reduced so that it approaches absolute zero (0 K or −273 °C), the electrical resistance disappears completely. The material is said to have become a **superconductor**. Its resistivity has dropped to zero and an electric current can pass through without transferring any energy to the conductor. The temperature at which the material becomes superconducting is known as the **critical temperature**, T_c.

The critical temperatures for metal superconductors are typically close to absolute zero, 1 to 4 K. Ceramic superconductors now exist that have critical temperatures as high as 125 K (−148 °C) (Fig 41).

Superconductors have important uses, for example carrying electrical power without losses, and constructing very strong electromagnets.

Essential Notes

At higher temperatures the ions of the semiconductor vibrate more. This would normally cause the resistance to rise. However, the release of conduction electrons is the dominant effect. This also explains the shape of the graph.

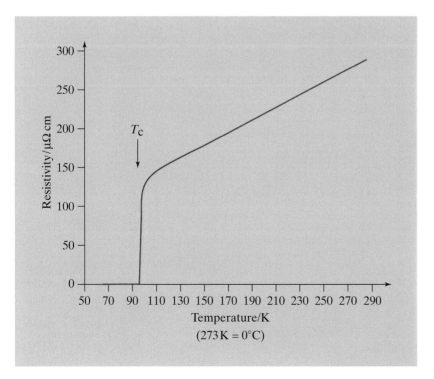

Fig 41
Resistivity against temperature for a 'high-temperature' superconductor

Circuits

Series resistor circuits

In series, the same current (I) flows through each resistor because, by conservation of charge, the current in any part of a series circuit is the same (Fig 42).

Fig 42
Three resistors in series

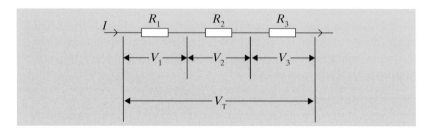

The p.d.s across the individual resistors add up to the total applied p.d.

$$V_T = V_1 + V_2 + V_3$$

Hence:

$$IR_T = IR_1 + IR_2 + IR_3$$

Dividing by I, the total resistance of any number of resistors is given by:

$$R_T = R_1 + R_2 + R_3\ldots$$

Parallel resistor circuits

When resistors are in parallel, the p.d. across each resistor is the same (Fig 43).

Fig 43
Three resistors in parallel

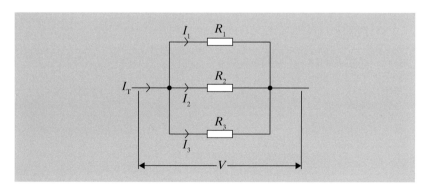

The total current is the sum of the currents through each resistor:

$$I_T = I_1 + I_2 + I_3$$

Hence:

$$\frac{V}{R_T} = \frac{V}{R_1} + \frac{V}{R_2} + \frac{V}{R_3}$$

Dividing by V, the total resistance of any number of resistors is given by:

$$\frac{1}{R_T} = \frac{1}{R_1} + \frac{1}{R_2} + \frac{1}{R_3}$$

Example

Draw diagrams to illustrate how three 10 Ω resistors can be connected in four different ways. Calculate the total resistance of each network of resistors.

Answer

$$R_T = 10 + 10 + 10 = 30\ \Omega$$

series $10 + 10 = 20\ \Omega$

parallel $\frac{1}{R_T} = \frac{1}{20} + \frac{1}{10} = \frac{1}{20} + \frac{2}{20} = \frac{3}{20}$ $\frac{R_T}{1} = \frac{20}{3} = 6.7\ \Omega$

parallel $\frac{1}{R} = \frac{1}{10} + \frac{1}{10} = \frac{2}{20}$ $R = \frac{10}{2} = 5\ \Omega$

series $R_T = 5 + 10 = 15\ \Omega$

$\frac{1}{R_T} = \frac{1}{10} + \frac{1}{10} + \frac{1}{10} = \frac{3}{10}$

$\frac{R_T}{1} = \frac{10}{3} = 3.3\ \Omega$

Energy and power in d.c. circuits

To make current flow, a p.d. must exist. The p.d. is the amount of electrical energy that must be transferred to the charge and is measured in joules per coulomb, or volts.

The charge releases the gained energy as it passes through components in a circuit (e.g. lamp, motor, resistor, etc.). All the potential energy lost by the charge is ultimately changed into heat.

Energy (W) is measured in joules.

Since

$$V = \frac{W}{Q} \text{ and } Q = It$$

then the energy converted to heat is given by:

energy change (work done) $W = VIt$

Power (P) is the rate of change of energy, and is measured in joules per second ($J\,s^{-1}$) or watts (W):

$$\text{power } P = VI$$

Thus, the energy delivered per second (power) by a 12 V battery supplying 2 A to a circuit is $24\,J\,s^{-1}$ or 24 W.

Example

Calculate the current supplied by a 1.5 V calculator battery with a power rating of 0.1 mW.

Answer

Rearrange $P = VI$ to $I = \dfrac{P}{V} = \dfrac{0.1 \times 10^{-3}}{1.5} = 0.07$ mA.

As current flows through resistors and lamps, heat is produced. The amount of power developed can be calculated using the equation:

$$P = VI$$

Essential Notes

The equation $P = I^2R$ is important because it shows that the heating effect is proportional to the square of the current. Therefore doubling the current will produce four times the rate of heating.

By substituting $V = IR$ into $P = VI$ we can arrive at an alternative equation:

$$P = I^2R$$

By substituting $I = \dfrac{V}{R}$ into $P = VI$ we can arrive at an alternative equation:

$$P = \dfrac{V^2}{R}$$

Example

(a) The power dissipated in a resistor R carrying a current I is P. If the resistance is doubled and the current halved, what power is now dissipated?

(b) A lamp is rated at 240 V, 60 W. What is its operating current?

(c) If the lamp operates at 200 V, what power is now dissipated by the lamp if its resistance remains the same when the voltage changes?

Answer

(a) Original power is given by $P = I^2R$, but $I_1 = \left(\dfrac{I}{2}\right)^2$ and $R_1 = 2R$

New power is given by $P_1 = \dfrac{I^2}{4} \times 2R = I^2\dfrac{R}{2}$ or $P_1 = \dfrac{1}{2}P$

(b) $P = VI$

Rearranging, $I = \dfrac{P}{V} = \dfrac{60}{240} = 0.25$ A

(c) $P = \dfrac{V^2}{R}$

Rearrange to find resistance of lamp $R = \dfrac{V^2}{P} = \dfrac{240 \times 240}{60} = 960\ \Omega$

Power of lamp $P = \dfrac{V^2}{R} = \dfrac{200 \times 200}{960} = 42\ \text{W}$

Conservation of charge and energy in circuits

In all circuits, electric charge is conserved, i.e. all the charge which arrives at a point must leave it. Current is a flow of charge, so this can be stated as follows.

Definition

At any point in a circuit where conductors join, the total current towards the point must equal the total current flowing away from the point.

or

The algebraic sum of currents at a junction is zero (Fig 44).

Essential Notes

This statement is known as Kirchhoff's First Law.

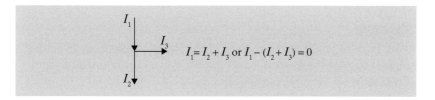

Fig 44
The sum of the currents at a junction is zero

In circuits, energy differences are expressed as potential differences and measured in volts. Energy is always conserved in all circuits. This results in the following rule.

Definition

The algebraic sum of potential differences around a closed circuit is zero.

Essential Notes

This statement is known as Kirchhoff's Second Law.

A closed circuit can contain a cell or battery – in this case the electromotive force (see p. 55) must equal the sum of the p.d.s around the circuit (Fig 45).

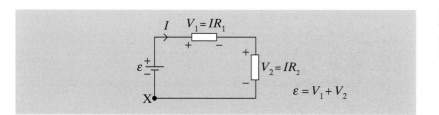

Fig 45
The electromotive force is the sum of the potential differences around the circuit

The algebraic sum of p.d.s around a closed circuit is zero. Therefore, going round the circuit in Fig 45 in a clockwise direction (with the current) starting and finishing at point X:

$$\varepsilon - V_1 - V_2 = 0$$

or:

$$\varepsilon = V_1 + V_2$$

If there are no electromotive forces in the identified closed loop then the sum of the p.d.s across individual components must equal zero (Fig 46).

Fig 46
When there are no electromotive forces in a closed circuit, the sum of the p.d.s across the individual components is zero

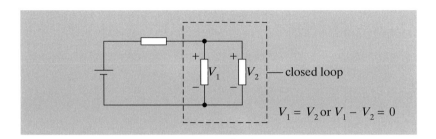

Example

Two 3 V, 0.3 A lamps are connected in parallel. Both lamps are to be run at their normal voltages from a 9 V supply by connecting a resistor in series to avoid damaging the lamps.

(a) Draw the diagram of this circuit.

(b) Calculate the current through the resistor.

(c) Calculate the voltage across the resistor.

(d) Calculate the value of the series resistor.

(e) The power rating of the resistor is 2.5 W. What is the maximum current allowed through the resistor?

(f) Comment on the suitability of using this resistor in this circuit.

Answer

(a)

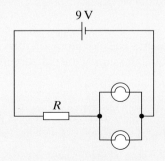

Fig 47

(b) Current through resistor = sum of currents through each lamp
= 0.3 + 0.3 = 0.6 A

(c) Voltage across resistor = $9 - 3 = 6$ V.

(d) $R = \dfrac{V}{I} = \dfrac{6}{0.6} = 10\ \Omega$

(e) Rearrange $P = I^2R$ to give maximum current:

$I^2 = \dfrac{P}{R} = \dfrac{2.5}{10} = 0.25\ \text{A}^2$

$I = \sqrt{0.25} = 0.5\ \text{A}$

(f) Current which flows will generate power in excess of the power rating of the resistor. Heat generated will 'burn out' or damage the resistor.

Potential divider

As its name suggests, a **potential divider** splits up the potential difference (voltage) from a source. This can be done using two or more resistors in series. In this case the p.d. across the output terminals varies according to the ratio of the resistances in series (Fig 48).

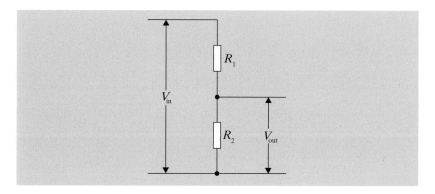

Fig 48
A potential divider

$$V_{\text{out}} = IR_2$$

$$V_{\text{in}} = I(R_1 + R_2)$$

The current I is the same in all parts of a series circuit, so:

$$\frac{V_{\text{out}}}{V_{\text{in}}} = \frac{IR_2}{I(R_1 + R_2)}$$

Therefore:

$$V_{\text{out}} = \frac{V_{\text{in}}R_2}{(R_1 + R_2)}$$

Examiners' Notes

When using this equation, take care to match up the resistor values with the labelled positions of R_1 and R_2.

See Fig 49 for an example of applying the potential divider equation.

Fig 49
Calculation of V_{out} with sample values

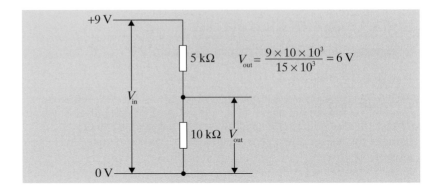

$$V_{out} = \frac{9 \times 10 \times 10^3}{15 \times 10^3} = 6 \text{ V}$$

Alternatively, a slider (variable) resistance may be used (Fig 50).

Fig 50
A slider

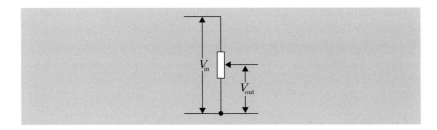

The p.d. across the output terminals varies with the position of the contact on the track (Fig 51).

Fig 51
Four examples showing variation of output voltage with contact position on a slider

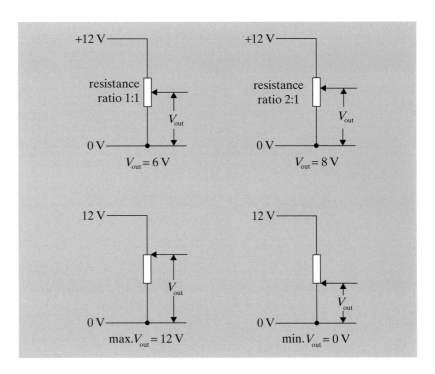

Example

(a) Calculate the potential difference between A and B in Fig 52.
(b) Calculate the potential difference between points A and B when another 5 Ω resistor is connected between them.

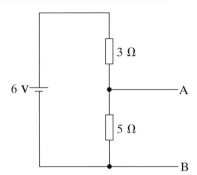

Fig 52

Answer

(a) Using the potential divider equation:

$$V_{out} = \frac{V_{in}R_2}{(R_1 + R_2)} = \frac{6 \times 5}{(3 + 5)} = 3.8 \text{ V}$$

(b)

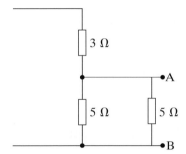

Fig 53

From Fig 53, the sum of two 5 Ω resistors in parallel is given by:

$$\frac{1}{R_T} = \frac{1}{5} + \frac{1}{5} = \frac{2}{5}$$

Therefore $R_T = \frac{5}{2} = 2.5 \text{ Ω}$

Applying the potential divider equation:

$$V_{out} = \frac{V_{in}R_T}{(R_1 + R_T)}$$

$$V_{out} = \frac{6 \times 2.5}{(3 + 2.5)} = 2.7 \text{ V}$$

Using sensors as part of potential dividers

Light and temperature sensors can be incorporated into potential dividers to vary the output voltage depending upon ambient conditions.

A light sensor (LDR) has a low resistance in bright light and a high resistance in darkness. Using the potential divider shown in Fig 54, the output voltage can be made to change with light intensity.

Fig 54
A light sensor as part of a potential
divider

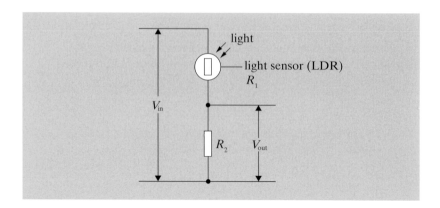

As the light gets dimmer the resistance of the LDR gets higher, the ratio of
resistances changes, and so does the voltage ratio, such that the output
voltage across R_2 decreases.

A temperature sensor (thermistor) has a low resistance when hot and a
high resistance when cold. Using the potential divider shown in Fig 55, the
output voltage can be made to change with temperature changes.

Fig 55
A temperature sensor as part of a
potential divider

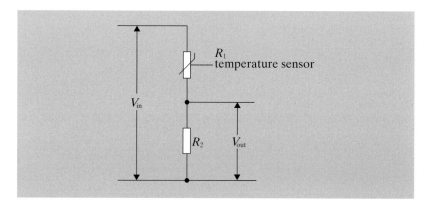

As the temperature gets colder the resistance of the thermistor increases
and the voltage ratio changes, such that the output across R_2 decreases.

If the values of the resistances of the light sensor or temperature sensor are
known, then the output of each potential divider under different
conditions can be calculated using the equation:

$$V_{out} = \frac{V_{in} R_2}{(R_1 + R_2)}$$

Example

A potential divider is made from a light sensor (whose resistance
decreases as the brightness of the light shining on it increases) and a
1000 Ω resistor (see Fig 56).

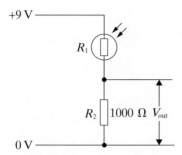

Fig 56

If the resistance of the light sensor is 500 ohms when in bright light, calculate the output voltage across R_2.

Answer

Using the equation $V_{out} = \dfrac{V_{in}\,R_2}{(R_1 + R_2)}$

$V_{in} = 9$ V, $R_1 = 500\ \Omega$ and $(R_1 + R_2) = (500 + 1000) = 1500\ \Omega$

$V_{out} = 9 \times \dfrac{1000}{1500} = 6$ V

Electromotive force and internal resistance

The **electromotive force** of a cell, or any other source of electrical energy (e.g. dynamo or thermocouple) can be defined as the p.d. across the source when no current flows, and is the energy per coulomb produced by the source. Electromotive force is usually shortened to **e.m.f.** and given the symbol ε.

$$\varepsilon = \frac{W}{Q}$$

Internal resistance

The materials inside a cell, or other power source, offer a resistance to the flow of current. This is known as the **internal resistance** of the cell (usually given the symbol r) and is measured in ohms (Fig 57).

- When no current flows in the circuit then the e.m.f. = p.d. across the cell.

- When current flows in the circuit there is a p.d. across $R\,(V_R = IR)$ and a p.d. across $r\,(V_r = Ir)$.

- Both energy and charge are conserved in the circuit.

- The current is the same in any part of the circuit.
  e.m.f. = sum of the p.d.s (Kirchhoff's second Law)

$$\varepsilon = IR + Ir = I(R + r)$$

Fig 57
Circuit showing the internal resistance of a cell

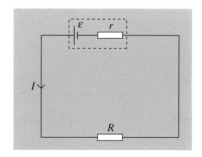

Essential Notes

The terminal p.d. of the cell is given by

$V = IR$

and not by $V = Ir$.

Example

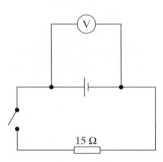

Fig 58

When the switch in the circuit in Fig 58 is open the voltmeter records 6.2 V. When the switch is closed the reading changes to 6 V.

(a) What causes the voltage across the battery to change?

(b) What is the value of the internal resistance of the battery?

(c) After the battery has been used for some time the internal resistance is 2.0 Ω. Calculate the current that flows when the switch is closed.

Answer

(a) When the switch is closed, a current, I, flows in the circuit. This current flows through the internal resistance, r, of the battery. This leads to a potential difference equal to Ir across the internal resistance, so that the voltage measured at the battery terminals is now less than when the switch was open.

(b) p.d. across internal resistance = ε − p.d. across external resistance
$$= 6.2 - 6 = 0.2\,\text{V}$$

Current through circuit $I = \dfrac{V}{R} = \dfrac{6}{15} = 0.4\,\text{A}$

Since it is a series circuit, there is also 0.4 A through r. So

$$r = \frac{V}{I} = \frac{0.2}{0.4} = 0.5\,\Omega$$

(c) $I = \dfrac{\varepsilon}{(R + r)} = \dfrac{6.2}{(15 + 2.0)} = 0.36\,\text{A}$

Measuring the e.m.f. and internal resistance of a cell

The circuit in Fig 59 may be used to measure the e.m.f. and the internal resistance of a cell.

Fig 59
Circuit for measuring the e.m.f. and internal resistance of a cell

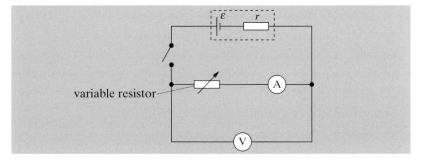

- The variable resistor is adjusted to enable a range of current (ammeter) and p.d. (voltmeter) readings to be recorded.

- A switch is used to break the circuit between readings to avoid 'running down' the cell.

- A graph is drawn (Fig 60), plotting p.d. on the y axis against current on the x axis.

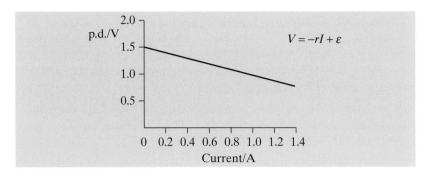

Fig 60
Graph of p.d. versus current through a variable resistor

- y axis intercept = e.m.f. (ε)
- $-$(gradient) = internal resistance, r

Example
Use the readings on the graph in Fig 60 to calculate:
(a) the e.m.f. of the cell
(b) the internal resistance of the cell.

Answer
(a) e.m.f. = intercept on y axis = 1.5 V

(b) internal resistance = $-$(gradient) = $\dfrac{1.5 - 0.7}{1.4} = 0.57\,\Omega$

Alternating currents

- The term **a.c.** can refer to **alternating current** and also alternating p.d.
- The term **d.c.** can refer to **direct current** and also direct p.d.
- a.c. can be represented as a sinusoidal graph varying between zero and a maximum value, either side of zero, in opposite directions (Fig 61).

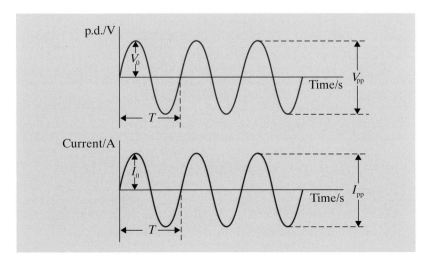

Fig 61
Alternating p.d. and current

Essential Notes

V_{pp} is peak-to-peak p.d.

I_{pp} is peak-to-peak current.

Several quantities can be calculated from graphs such as those in Fig 61, including root mean square value, peak value, peak-to-peak value, time period and frequency.

The **root mean square value (r.m.s.)** – since the a.c. current or p.d. is continually changing in value it is impossible to assign a fixed value over a number of cycles (the mean would be zero). The r.m.s. current produces the same heating effect in a resistor as the equivalent d.c.; i.e. I_{rms} produces the same heating effects as $I_{d.c.}$

Essential Notes

In a practical situation it is better to measure V_{pp} and divide by 2 in order to obtain a value for V_0 since there is less error in measuring a longer length.

The **peak value** of an a.c. current or p.d. is the maximum displacement from the zero line in either direction and is labelled either I_0 or V_0 respectively on the graph.

To convert peak values to r.m.s. values the following equations can be used:

$$I_{rms} = \frac{I_0}{\sqrt{2}} \qquad V_{rms} = \frac{V_0}{\sqrt{2}}$$

The **peak-to-peak value** of an a.c. current or p.d. is the maximum displacement across both directions and is labelled either I_{pp} or V_{pp} respectively on the graph.

The **time period** (T) of an a.c. current or p.d. is the time taken for one complete cycle. The unit is seconds or, more usually, ms or µs.

The **frequency** (f) of an a.c. current or p.d. is the number of complete cycles per second. The unit is hertz (Hz).

Period and frequency are linked by the equation:

$$T = \frac{1}{f}$$

Oscilloscope

Fig 62
A typical oscilloscope

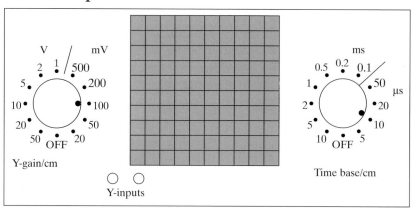

An **oscilloscope** is a device for displaying waveforms. It can be used to:

• measure a.c. and d.c. voltages

• measure small time intervals

• measure frequencies of alternating currents and voltages.

Interpreting the oscilloscope screen

The screen of an oscilloscope is made from a fluorescent material, so that when an electron hits it, the screen lights up as a bright spot. Two sets of plates on either side of the electron beam can make the spot move horizontally or vertically, depending on the application of voltages to the **X-** or **Y-plates**.

By applying a varying p.d. to the X-plates, the spot can be made to move horizontally at a constant speed across the screen from left to right only. This is known as the **time base**. It is usually set in milliseconds (ms) or microseconds (μs) per division on the screen. Screen divisions are usually 1 cm. The X-plate controls are used to set the time base when measuring small time intervals and frequencies.

When no external voltage is applied, the oscilloscope screen appears as in Fig 63.

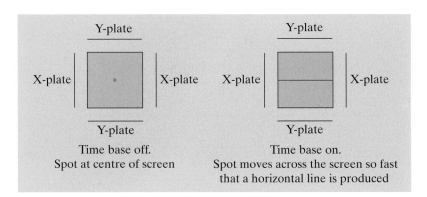

Fig 63
Oscilloscope screen with no external voltage applied

By applying an external p.d. to the Y-plates (either d.c. or a.c. depending on which is being measured), the spot can be made to move vertically. The larger the p.d., the greater the deflection recorded on the screen.

The amount of deflection is also controlled by the Y-gain. Oscilloscopes can amplify the input voltage, allowing them to display small inputs more clearly. The Y-gain is usually set in volts per division on the screen.

When an external voltage is applied, the oscilloscope screen appears as in Fig 64.

Measuring a.c./d.c. voltage

When setting up an oscilloscope:

- switch the time base to 'off'

- centralise the spot on the screen

- choose a suitable sensitivity for the Y-gain, e.g. 1 V per division

- connect the voltage to be measured to the Y-inputs of the oscilloscope.

Fig 64
Oscilloscope screen with an external p.d. applied

d.c. voltage applied to Y-plates	a.c. voltage applied to Y-plates
Time base off, positive d.c. voltage applied to upper Y-plate. Spot moves towards upper plate. Applied voltage equal to displacement of spot.	Time base off, a.c. voltage applied to Y-plates. Spot moves up and down in a straight line. Height of line equals peak-to-peak voltage.
Time base off, negative d.c. voltage applied to upper Y-plate Spot moves towards lower plate. Applied voltage equal to displacement of spot.	Time base on, a.c. voltage applied to Y-plates. Spot moves up and down and across the screen. Number of cycles displayed depends on time base setting.

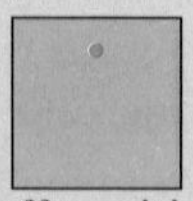

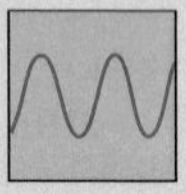

When measuring a d.c. voltage:

- watch for the deflection of the spot (Fig 65)

- adjust the sensitivity to produce maximum displacement on the screen

- calculate the input voltage value using the equation:

$$V_{in} = \text{spot deflection (in divisions)} \times \text{Y-gain setting (in volts/division)}$$

Fig 65
Measuring d.c. voltage

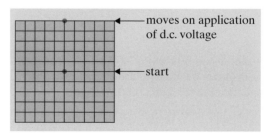

Deflection 5 divisions

Y-gain = 0.5 V per division

$V_{in} = 5 \times 0.5 = 2.5\,V$

When measuring an a.c. voltage:

- watch the vertical line which appears (Fig 66)

- adjust the sensitivity to ensure that the line fits on the screen, with maximum height

- calculate the peak-to-peak voltage using the equation:

$$V_{pp} = \text{height of vertical line (in divisions)} \times \text{Y-gain setting (in volts/division)}$$

- The peak value is:

$$V_0 = \frac{V_{pp}}{2}$$

Height of line = 6 divisions

Y-gain = 3 V/division

$V_{pp} = 6 \times 3 = 18$ V

$V_0 = \dfrac{18}{2} = 9$ V

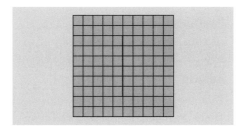

Fig 66
Measuring a.c. voltage

Essential Notes

V_{rms} value can be calculated from the peak value:

$$V_{rms} = \frac{V_0}{\sqrt{2}}$$

$$V_{rms} = \frac{9}{\sqrt{2}} = 6.4 \text{ V}$$

Measuring the period and frequency of an a.c. voltage waveform

To measure the period and frequency of an a.c. voltage waveform:

- switch on the time base and adjust until a horizontal line appears

- centre the line on the screen

- connect the input signal to the Y-plates of the oscilloscope

- watch for the trace to appear on the screen

- adjust the time base setting in order to observe a convenient number of complete cycles

- measure the distance along the screen for, say, 3 cycles (Fig 67)

- calculate the distance for one full cycle by dividing the above answer by 3

- calculate the periodic time (T) using the equation:

> period (s) = length of single cycle (in divisions)
> × time base value (in s/divisions)

- frequency can then be calculated using:

$$f = \frac{1}{T}$$

Length for 3 cycles = 4.8 divisions

Length of one cycle = 1.6 divisions

Time base set at 2 ms/division

Period T (ms) = 1.6 × 2 ms/division = 3.2 ms

Frequency $= \dfrac{1}{T} = \dfrac{1}{3.2 \times 10^{-3} \text{s}} = 313$ Hz

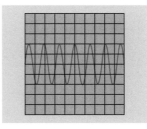

Fig 67
Measuring period and frequency of a.c.

Measuring small time intervals

Small time intervals can be measured using the time base setting. For example, measuring heart rate, or the time for a wave to be reflected in a solid metal bar or tube.

Fig 68
Oscilloscope trace of a heartbeat

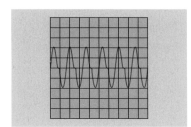

Example

Use the oscilloscope trace in Fig 68 to calculate the heartbeat rate per minute. The time base setting is 0.4 s/division.

Answer

Time base setting = 0.4 s/division

There are 6 heartbeats in 10 divisions

Length of one beat = 1.67 divisions

Period = 1.67 × 0.4 = 0.67 s

Frequency = $\dfrac{1}{0.67}$ = 1.5 Hz or 90 beats per min

Example

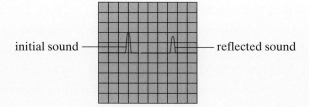

initial sound — reflected sound

Fig 69

A sound and its reflection travelling through an iron bar and back are picked up by a microphone and displayed on an oscilloscope (Fig 69). If the time base is set at 2 ms/division, how long does the sound take to travel down the bar and back?

Answer

Time base = 2 ms/division

There are 4.5 divisions between pulses.

Total time elapsed = $2 \times 10^{-3} \times 4.5$ = 9 ms

Example

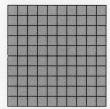

Fig 70

Fig 70 shows an oscilloscope screen with a line running centrally across the screen. The time base is set to 2 ms/division and the Y-gain setting is 2 V/division.

Draw the trace seen on the screen when a 6 V_{rms} a.c. signal of frequency 200 Hz is applied to the Y-inputs.

Answer

Calculating number of cycles on screen.

Frequency = 200 Hz

Period $= \dfrac{1}{f} = \dfrac{1}{200} = 5 \times 10^{-3}$ s or 5 ms

Time base is set at 2 ms/division and there are ten divisions on the screen. Therefore there will be 4 cycles on the screen.

Calculating the peak voltage.

$V_{\text{peak}} = V_{\text{rms}} \times \sqrt{2} = 6 \times \sqrt{2} = 8.5$ V

Y-gain is set at 2 V/division so amplitude needs to be 4.25 divisions either side of central horizontal line.

Therefore the trace produced would be as shown in Fig 71.

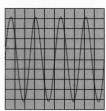

Fig 71

Examination preparation

How Science Works

Data and formulae

Practice exam-style questions

Answers, explanations, hints and tips

How Science Works

As well as understanding the physics in this unit, you are expected to develop an appreciation of the nature of science, the way that scientific progress is made and the implications that science has for society in general. GCSE and A-level science syllabuses refer to these areas as '*How Science Works*'.

The *How Science Works* element of your course, which also includes important ideas about experimental physics, may be assessed in the written examination papers as well as in the internally assessed unit, the Investigative Skills Assessment or ISA. The concepts included in *How Science Works* may be divided into several areas.

Theories and models

Physicists use theories and models to attempt to explain their observations of the universe around us. These theories or models of the real world can then be tested against experimental results. Scientific progress is made when experimental evidence is found that supports a new theory or model.

You are expected to be aware of historical examples of how scientific theories and models have developed and how this has changed our knowledge and understanding of the physical world. Examples include Dirac's prediction of antimatter. His theoretical equation which described the electron also predicted the existence of its antiparticle – the positron. This was later vindicated when Anderson designed an experiment that showed positron tracks in a cloud chamber. Similarly, Pauli postulated, by applying the physical principles of the conservation of energy and the conservation of momentum, that there must be a previously unobserved particle emitted with the electron in beta decay. Eventually, many years later, experimental results showed that he was right and the neutrino theory became generally accepted.

Experimental results may of course disprove a theory that was previously accepted. Rutherford's scattering experiment showed that the 'plum-pudding' model of an atom could not be correct. Rutherford's results led to a new model, that of the nuclear atom.

You should know the meaning of the terms 'hypothesis' and 'prediction'.

- A **hypothesis** is a tentative idea or theory, or explanation of an observation.

- A **prediction** from a hypothesis or theory is a forecast that can be tested by experiment.

If a reliable experiment does not support a hypothesis, then the hypothesis is likely to be abandoned or modified. Hypotheses are not usually widely accepted until the experimental results have been repeated by a number of independent scientists. It may take many experimental tests until a set of hypotheses become accepted as a scientific theory (see the figure below). Even then, a scientific theory is always capable of being later refuted if compelling experimental evidence suggests that a new explanation is necessary.

The stages of scientific research

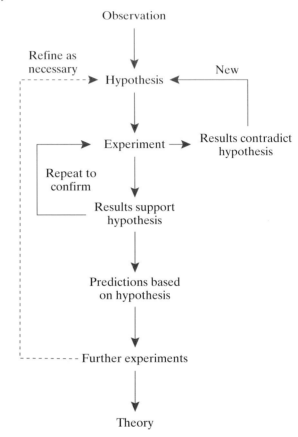

Experimental techniques

You are expected to develop the skills of experimental planning, observation, recording and analysis. These skills will mainly be assessed in the practical coursework, the ISA, or by the externally-marked practical assignment, the EMPA. This section contains some general advice for carrying out experimental work in physics.

When you plan an experiment you need to be able to identify the dependent, independent, and control variables that are involved.

- The **independent variable** is the physical quantity that you deliberately change.

- The **dependent variable** changes as a result of this.

For example, if you are asked to investigate how the length of a piece of metal wire affects its electrical resistance, the length is the independent variable, and the electrical resistance is the dependent variable. Any other variables that may have an impact on the outcome need to be controlled so that the conclusions of the experiment are clear. These are known as the **control variables**. In the example of the wire, two of the control variables are the cross-sectional area of the wire and its temperature.

You will need to select appropriate apparatus, including measuring instruments of a suitable precision and accuracy. These two terms are often confused.

- **Accuracy** refers to how close the reading is to the accepted value.
- **Precision** refers to the number of significant figures that the measurement is made to.

For example, an electronic balance that gives an answer to 0.01 g, e.g. 3.24 g, is capable of more precise measurements than a balance that measures in grams only, e.g. 3 g. If several readings of the same measurement are closely grouped together, the readings are said to be **precise**. If the readings agree with a known mass, they are said to be **accurate**. The analogy of rifle shots at a target may be used to differentiate between accuracy and precision (see the figure below).

Figure A shows accurate shooting, since the bullets (or readings) are close to the centre of the target. But the shooting is not precise, since the bullets are widely scattered. Figure B shows precise shooting, since the bullets (readings) are closely grouped, but not accurate, since the bullets (readings) are not close to the centre of the target (accepted value).

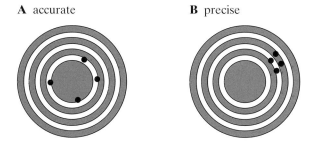

A accurate   **B** precise

When choosing your apparatus you should be aware that ICT can be used to assist with the collection and analysis of experimental data. This may mean using a suitable sensor, attached to a data logger to take the readings and then a spreadsheet to help to analyse them. For example, suppose that you wanted to investigate the current surge that passes through the filament of an incandescent light bulb when it is first turned on. The light bulb reaches its operating temperature very quickly, so the readings need to be taken rapidly. A current sensor attached to a data logger could take the readings at the required rate. Conversely, sometimes the readings need to be taken over a long time period, for example if you wanted to investigate the cooling of a house overnight. A temperature sensor and a data logger would make the job somewhat less tedious!

Planning an experiment also means being aware of any risks to health and safety, and taking any necessary precautions. For example, you need to wear protective goggles when stretching a metal wire.

You also need to plan to reduce experimental errors. There are two types of error: systematic and random.

- **Systematic errors** cannot be reduced by repeating the measurement; for example, using an electronic balance which is not zeroed would lead to a systematic error.

- **Random errors** occur when taking readings, such as the timing errors when measuring the period of a pendulum. These can be reduced by repeating the readings and finding the mean, since the errors are random and may fluctuate above and below an average reading.

It is important to identify the percentage uncertainty associated with each reading. For example, if a length is measured using a ruler with millimetre divisions the reading in centimetres may be given to a precision of ± 0.1 cm. A reading of 25.2 ± 0.1 cm has a percentage uncertainty of $(^{0.1}/_{25.2}) \times 100\% = 0.4\%$.

The percentage uncertainty can be reduced by using a more precise measuring device, such as a micrometer or vernier callipers instead of a ruler. The percentage uncertainty can also be reduced by increasing the size of the quantity to be measured. For example, to measure the thickness of a sheet of paper, you could measure the thickness of 100 sheets of paper, and then divide by 100 to find the value for a single sheet. When timing the period of a pendulum it is good practice to time a number of oscillations, say 10, and then divide by 10.

When you tabulate your readings you should ensure that the columns are headed with the quantity *and* the unit that it is measured in. It is good practice to also include the uncertainty associated with that reading in the heading, as in the table here.

Length/m ± 0.001 m	Time/s ± 0.1 s
0.023	2.3
0.031	3.1
0.04*	4.0

* see below

You should always quote figures in your results table to the appropriate degree of precision, and be consistent. The length reading in the last row of the table isn't correct – it should read 0.040.

The most significant uncertainty in the readings taken in an experiment may determine whether you can draw a reliable conclusion or not. You can also test reliability by repeating the experiment a number of times and comparing your results. In practice scientists share their results through publication. Reliability is tested by other scientists who try to replicate the work, or check the results using a different experimental method.

When you plot your results on a graph, you should choose a scale so that the range of your points covers at least half of the graph paper, in both the *x*- and *y*-directions. Choose scales that have divisions that are easy to interpret accurately, i.e. divisions in multiples of 2, 5 or 10 rather than 3 or 7.

You must label each axis of the graph with the quantity and the unit; conventionally this is done using a solidus, for example length/m. The best-fit line or curve should be drawn so as to minimise the total distance of points from the line. In practice this will mean that some points fall

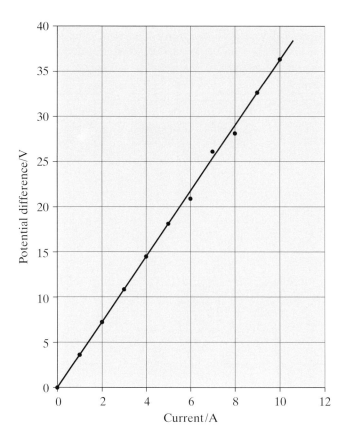

Graph to show how the potential difference across a wire affects the current through it

above the line and some below (see the graph here). Any results that fall outside the expected range of values (anomalous values) can be ignored when choosing the best line, but you should identify these anomalous readings on your graph. You should also repeat these readings whenever possible and try to explain why they do not follow the trend. Finally, when measuring the gradient of a line, choose a large section of the graph so as to reduce errors.

Applications and implications of science

Scientific advances have greatly improved the quality of life for the majority of people, and developments in technology, medicine and materials continue to further these improvements at an increasing rate. However, technologies themselves pose significant risks that have to be balanced against the benefits. For example, the way in which electrical energy is generated poses many questions. Nuclear power is capable of generating large amounts of energy and does not emit carbon dioxide, but it does produce radioactive waste which needs to be stored safely for thousands of years.

Scientific findings and technologies enable advances to be made that have potential benefit for humans; however, the scientific evidence available to policy makers may be incomplete. Political decision-makers are influenced by many things, including prior beliefs, vested interests, public opinion and the media, as well as by expert scientific evidence.

Data and formulae

FUNDAMENTAL CONSTANTS AND VALUES

Quantity	Symbol	Value	Units
speed of light in vacuo	c	3.00×10^8	$\mathrm{m\,s^{-1}}$
permeability of free space	μ_0	$4\pi \times 10^{-7}$	$\mathrm{H\,m^{-1}}$
permittivity of free space	ε_0	8.85×10^{-12}	$\mathrm{F\,m^{-1}}$
charge of electron	e	-1.60×10^{-19}	C
the Planck constant	h	6.63×10^{-34}	J s
gravitational constant	G	6.67×10^{-11}	$\mathrm{N\,m^2\,kg^{-2}}$
the Avogadro constant	N_A	6.02×10^{23}	$\mathrm{mol^{-1}}$
molar gas constant	R	8.31	$\mathrm{J\,K^{-1}\,mol^{-1}}$
the Boltzmann constant	k	1.38×10^{-23}	$\mathrm{J\,K^{-1}}$
the Stefan constant	σ	5.67×10^{-8}	$\mathrm{W\,m^{-2}\,K^{-4}}$
the Wien constant	α	2.90×10^{-3}	m K
electron rest mass (equivalent to $5.5 \times 10^{-4}\,\mathrm{u}$)	m_e	9.11×10^{-31}	kg
electron charge–mass ratio	e/m_e	1.76×10^{11}	$\mathrm{C\,kg^{-1}}$
proton rest mass (equivalent to $1.00728\,\mathrm{u}$)	m_p	1.67×10^{-27}	kg
proton charge–mass ratio	e/m_p	9.58×10^7	$\mathrm{C\,kg^{-1}}$
neturon rest mass (equivalent to $1.00867\,\mathrm{u}$)	m_n	1.67×10^{-27}	kg
gravitational field strength	g	9.81	$\mathrm{N\,kg^{-1}}$
acceleration due to gravity	g	9.81	$\mathrm{m\,s^{-2}}$
atomic mass unit (1 u is equivalent to 931.3 MeV)	u	1.661×10^{-27}	kg

ASTRONOMICAL DATA

Body	Mass/kg	Mean radius/m
Sun	2.0×10^{30}	7.0×10^8
Earth	6.0×10^{24}	6.4×10^6

GEOMETRICAL EQUATIONS

arc length $= r\theta$

circumference of circle $= 2\pi r$

area of circle $= \pi r^2$

area of cylinder $= 2\pi rh$

volume of cylinder $= \pi r^2 h$

area of sphere $= 4\pi r^2$

volume of sphere $= \dfrac{4}{3}\pi r^3$

PARTICLE PHYSICS

Rest energy values

Class	Name	Symbol	Rest energy /MeV
photon	photon	γ	≈ 0
lepton	neutrino	ν_e	≈ 0
		ν_μ	≈ 0
	electron	$e^\pm$	0.510999
	muon	$\mu^\pm$	105.659
mesons	pion	$\pi^\pm$	139.576
		π^0	134.972
	kaon	$K^\pm$	493.821
		K^0	497.762
baryons	proton	p	938.257
	neutron	n	939.551

Properties of quarks (antiquarks have opposite signs)

Type	Charge	Baryon number	Strangeness
u	$+\frac{2}{3}$	$+\frac{1}{3}$	0
d	$-\frac{1}{3}$	$+\frac{1}{3}$	0
s	$-\frac{1}{3}$	$+\frac{1}{3}$	-1

Properties of leptons

	lepton number
particles: e^-, ν_e; μ^-, ν_μ	$+1$
antiparticles: e^+, $\bar{\nu}_e$; μ^-, $\bar{\nu}_\mu$	-1

Photons and energy levels

photon energy	$E = hf = hc/\lambda$
photoelectricity	$hf = \phi + E_{k(max)}$
energy levels	$hf = E_1 - E_2$
de Broglie wavelength	$\lambda = \dfrac{h}{p} = \dfrac{h}{mv}$

ELECTRICITY

current and p.d.	$I = \dfrac{\Delta Q}{\Delta t}$ $V = \dfrac{W}{Q}$ $R = \dfrac{V}{I}$
e.m.f.	$\varepsilon = \dfrac{E}{Q}$ $\varepsilon = I(R + r)$
resistors in series	$R = R_1 + R_2 + R_3 + \dots$
resistors in parallel	$\dfrac{1}{R} = \dfrac{1}{R_1} + \dfrac{1}{R_2} + \dfrac{1}{R_3} + \dots$
resistivity	$\rho = \dfrac{RA}{L}$
power	$P = VI = I^2R = \dfrac{V^2}{R}$
alternating current	$I_{rms} = \dfrac{I_0}{\sqrt{2}}$ $V_{rms} = \dfrac{V_0}{\sqrt{2}}$

MECHANICS

moments

moment $= Fd$

velocity and acceleration $v = \dfrac{\Delta s}{\Delta t}$ $a = \dfrac{\Delta v}{\Delta t}$

equations of motion

$v = u + at$ $s = \dfrac{(u + v)}{2}t$

$v^2 = u^2 + 2as$ $s = ut + \dfrac{at^2}{2}$

force $F = ma$

work, energy and power $W = Fs \cos \theta$

$E_k = \tfrac{1}{2}mv^2$ $\Delta E_p = mg\Delta h$

$P = \dfrac{\Delta W}{\Delta t}$ $P = Fv$

efficiency $= \dfrac{\text{useful output power}}{\text{input power}}$

MATERIALS

density $\rho = \dfrac{m}{V}$ Hooke's Law $F = k\Delta L$

Young modulus $= \dfrac{\text{tensile stress}}{\text{tensile strain}}$ tensile stress $= \dfrac{F}{A}$

tensile strain $= \dfrac{\Delta L}{L}$

energy stored $E = \tfrac{1}{2}F\Delta L$

WAVES

wave speed $c = f\lambda$ period $T = \dfrac{1}{f}$

fringe spacing $w = \dfrac{\lambda D}{s}$ diffraction grating $d\sin\theta = n\lambda$

refractive index of a substance s $n = \dfrac{c}{c_s}$

for two different substances of refractive indices n_1 and n_2,

law of refraction $n_1 \sin\theta_1 = n_2 \sin\theta_2$

critical angle $\sin\theta_c = \dfrac{n_2}{n_1}$ for $n_1 > n_2$

Practice exam-style questions

1 (a) Radium-228, $^{228}_{88}$Ra, is a radioactive isotope of the element radium. Explain what *isotopes* are.

_____

_____ 2 marks

   (b) How many protons, neutrons and electrons are there in an atom of $^{228}_{88}$Ra?

   Protons _____

   Neutrons _____

   Electrons _____   3 marks

   (c) Radium-228 decays into radon, symbol Rn, by emitting an alpha particle. Write an equation for this decay.

   $^{228}_{88}$Ra $\rightarrow$ _____ 3 marks

   (d) What is the specific charge of the alpha particle?

_____

_____

_____ 3 marks

                                                        Total Marks: 11

2 Carbon-14 emits beta particles. The decay is written as

$$^{14}_{6}C \rightarrow {}^{14}_{7}N + e^- + \bar{\nu}_e$$

   (a) What particle is represented by the symbol $\bar{\nu}_e$ in the equation above?

_____ 1 mark

   (b) Pauli was the first person to suggest that this particle might exist, even though it had never been observed. Suggest one reason why he did so.

_____ 1 mark

   (c) When a beta particle is emitted from a carbon-14 nucleus, one of the neutrons in the nucleus decays to a proton. This decay is caused by the weak interaction. A Feynman diagram can be used to represent this decay. Complete the Feynman diagram below which shows beta decay.

                                                                    4 marks

(d) Neutrons and protons are composed of quarks. Describe the quark structure of the neutron and explain what happens to it during beta decay.

_____

_____

_____ 3 marks

Total Marks: 9

3   A bubble chamber is a device which allows us to see the tracks made by high-energy particles, like electrons or protons. The diagram below is a sketch of a bubble chamber photograph and shows the phenomenon of pair production.

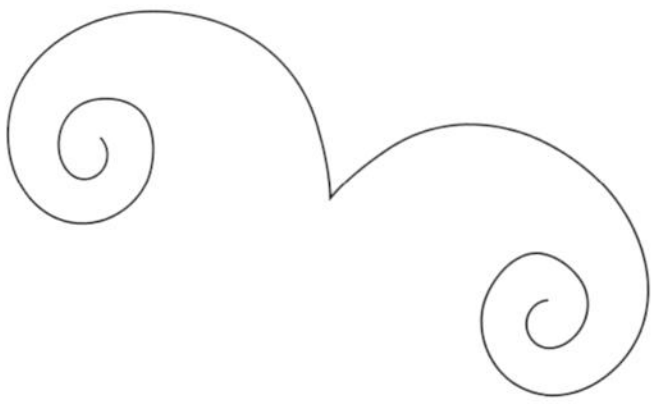

(a) Explain what is meant by the term *pair production*.

_____

_____

_____ 3 marks

(b) Explain why the tracks in the diagram are shaped as they are.

_____

_____

_____ 3 marks

Total Marks: 6

4   Neon signs are used for advertising. They are glass tubes that contain neon gas at a low pressure. When an electric current is passed though the neon gas it emits light. Atoms in the neon gas are excited or ionised by collision with electrons that move through the tube.

(a) Explain what is meant by an atom becoming *excited*.

_____

_____ 1 mark

(b) Explain what is meant by an atom becoming *ionised*.

_____

_____ 1 mark

(c) The electrons in atoms can only orbit in certain energy levels. Here is part of the energy level diagram for a hydrogen atom.

level 4 —————————————— −0.54 eV
level 3 —————————————— −0.85 eV

level 2 —————————————— −1.51 eV

level 1 —————————————— −3.4 eV

ground state —————————————— −13.6 eV

When an electron moves to a lower energy level it emits a photon of light.

(i) Calculate the energy lost by an electron as it moves from level 3 to the ground state. Express your answer in joules.

_____

_____ 2 marks

(ii) What is the energy of the photon that is emitted when an electron moves from level 3 to the ground state?

_____ 1 mark

(iii) Calculate the wavelength of light that is emitted when an electron moves from level 3 to the ground state.

_____

_____ 2 marks

Total Marks: 7

5 The photoelectric effect occurs when light shines onto a metal surface and causes electrons to be emitted from the surface. Einstein's photoelectric equation describes the process:

$$hf = \phi + E_k$$

(a) State *and* explain what is meant by each of the terms in the equation above.

    (i) hf _____

    _____ 2 marks

    (ii) ϕ _____

    _____ 2 marks

    (iii) E_k _____

    _____ 2 marks

(b) The value of ϕ for a clean zinc surface is 4.3 eV. What is the longest wavelength of light that could cause photoemission from the zinc?

_____

_____

_____ 3 marks

Total Marks: 9

6 An incandescent electric light bulb has a thin metal wire, known as a filament, which glows white hot when a current passes through it. When it is connected to a potential difference of 230 V and it has reached its working temperature, the electric current through the filament is 0.50 A.

(a) Explain what is meant by *electric current*.

_____

_____ 2 marks

(b) Potential difference is measured in volts. Give a definition of the volt.

_____

_____ 2 marks

(c) Calculate the resistance of the light bulb filament at its working temperature.

_____

_____

_____ 3 marks

(d) The light bulb is switched on all day, for 24 hours. How much electrical energy would it transfer (to heat and light) in this time?

_____

_____

_____

_____ 4 marks

(e) For a short time, when the bulb is first switched on, the current is much higher than 0.5 A. Suggest why this is.

_____

_____

_____ 3 marks

Total Marks: 14

7 A student is asked to find the resistivity, ρ, of nichrome, which is a metal alloy. The student is given a 2 m length of nichrome wire which has a diameter of approximately 0.5 mm.

Describe how the student could find the resistivity. Explain what measurements need to be taken and what measuring instruments would be used. Include a suitable circuit diagram. Explain what steps the student could take to get an answer which is accurate and reliable.

(You will be assessed on your ability to communicate your answer clearly.)

_____

_____

_____

_____

_____

Total Marks: 7

8 An oscilloscope is used to measure an alternating current. The screen shows the following wave.

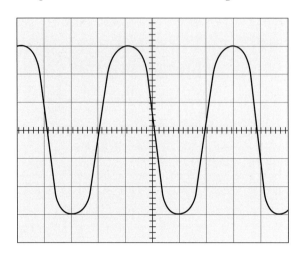

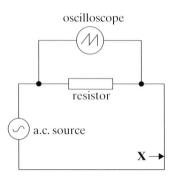

The time base is set at 5 ms per division.

The Y-sensitivity is set at 2 V per division.

(a) Calculate the peak voltage of the signal.

_____

_____ 2 marks

(b) Calculate the frequency of the signal.

_____

_____

_____ 3 marks

(c) A diode is placed in the circuit at point X. Sketch the trace that you might see on the oscilloscope.

2 marks

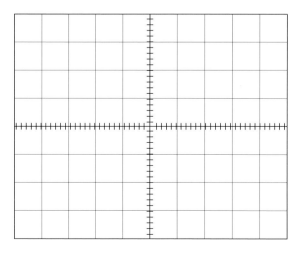

Total Marks: 7

Answers, explanations, hints and tips

Question	Answer		Marks
1 (a)	Isotopes are different forms of an atom (of the same element) which have the same number of protons in the nucleus but a different number of neutrons	(1) (1)	2
1 (b)	Protons 88 Neutrons 140 Electrons 88 (must be equal to the number of protons)	(1) (1) (1)	3
1 (c)	$\rightarrow {}^{4}_{2}\text{He} + {}^{224}_{86}\text{Rn}$ Correct symbol for alpha, either ${}^{4}_{2}\text{He}$ or ${}^{4}_{2}\alpha$ Correct proton number for Rn = 86 Correct nucleon number for Rn = 224	 (1) (1) (1)	3
1 (d)	Mass of alpha = 2 × mass of proton + 2 × mass of neutron = 6.644×10^{-27} kg Charge of alpha = twice charge of proton = $2 \times 1.6 \times 10^{-19}$ C = 3.2×10^{-19} C Charge–mass ratio = 3.2×10^{-19} / 6.644×10^{-27} = 4.82×10^{7} C kg^{-1} (must include the correct unit)	(1) (1) (1)	3
			Total 11
2 (a)	Antineutrino	(1)	1
2 (b)	So as to conserve energy / momentum	(1)	1
2 (c)		(4)	4
2 (d)	A neutron is composed of 3 quarks These are up, down, down (udd) During beta decay, one of the down quarks is transformed to an up	(1) (1) (1)	3
			Total 9
3 (a)	Pair production is when two particles one of matter and one of antimatter are produced from energy / a gamma-ray photon	(1) (1) (1)	3
3 (b)	The particle tracks are curved due to a magnetic field They curve in opposite directions because the particles have opposite charge They curve with the same radius because the particles have the same charge–mass ratio / same mass / equal but opposite charge The tracks spiral as the particles slow down	(1) (1) (1) (1) (any 3)	3
			Total 6

Question	Answer		Marks
4 (a)	Excited means that an electron in the atom moves to a higher energy level (due to a collision with the electron)	(1)	1
4 (b)	Ionised means that an atomic electron is removed from the atom / knocked out of orbit (due to a collision with the electron)	(1)	1
4 (c)(i)	$-0.85 - (-13.6) = 12.75$ eV	(1)	
	$= 2.04 \times 10^{-18}$ J	(1)	2
(ii)	12.75 eV $= 2.04 \times 10^{-18}$ J (as in i above)	(1)	1
(iii)	$E = hc/\lambda$ so $\lambda = hc/E$	(1)	
	$= 6.63 \times 10^{-34} \times 3 \times 10^8 / 2.04 \times 10^{-18} = 9.75 \times 10^{-8}$ m	(1)	2
			Total 7
5 (a)(i)	h = Planck constant, f = frequency of light	(1)	
	hf is the energy of the incident photon	(1)	2
(ii)	ϕ = the work function	(1)	
	which is the minimum energy required to cause photoemission for a given metal	(1)	2
(iii)	E_k is the maximum kinetic energy	(1)	
	of the emitted electrons	(1)	2
5 (b)	$\phi = E_{min} = hc/\lambda_{max}$	(1)	
	so $\lambda_{max} = hc/\phi$	(1)	
	$= 6.63 \times 10^{-34} \times 3 \times 10^8 / (4.3 \times 1.6 \times 10^{-19}) = 2.89 \times 10^{-7}$ m	(1)	3
			Total 9
6 (a)	Electric current is defined as the rate	(1)	
	of flow of charge	(1)	
	(or the charge that flows past a point in a given time)		2
6 (b)	One volt is the potential difference between two points when		
	one coulomb of charge	(1)	
	transfers one joule of energy as it moves between them	(1)	
	(1 volt = 1 joule per coulomb)		2
6 (c)	Resistance $R = V/I$	(1)	
	$= 230/0.5 = 460$ Ω	(2)	
	(one mark is for the correct unit, Ω)		3
6 (d)	Power $P = IV = 0.5 \times 230 = 115$ W; which is 115 joules per second	(1)	
	There are $24 \times 60 \times 60 = 86\,400$ seconds in a day	(1)	
	so the energy transferred is $115 \times 86\,400 = 9.94 \times 10^6$ J	(2)	
	(one mark is for the correct unit)		4
6 (e)	Filament / bulb is cold at first	(1)	
	so its resistance is lower	(1)	
	so more current flows, since $I = V/R$	(1)	3
			Total 14

Question	Answer		Marks
7	Circuit diagram: includes the wire, a power supply (variable), an ammeter in series with the wire and a voltmeter in parallel with it	(1)	1
	Measurements:		
	Measure the voltage and current	(1)	
	Use these to find the resistance of the wire	(1)	
	Measure the length of the wire with a metre ruler or tape	(1)	
	Measure the diameter of the wire with a micrometer	(1)	
	Use the equation $\rho = AR/l$ to find ρ	(1)	
	(at least 3 of the above)		3
	To get an accurate answer:		
	Measure diameter several times in different places / directions and find the mean	(1)	
	Use a long piece of the wire	(1)	
	Use a low current to keep the temperature of the wire down	(1)	
	Take several sets of readings for different lengths l and find mean value of ρ	(1)	
	Plot a graph of R vs l and find ρ from gradient	(1)	
	(at least 3 of the above)		3
			Total 7
8 (a)	Height of trace from x-axis is 3 divisions	(1)	
	Peak voltage is $3 \times 2 = 6$ V	(1)	2
8 (b)	One cycle is about 4 divisions (allow down to 3.7)	(1)	
	Time period $= 5$ ms $\times 4 = 20$ ms	(1)	
	Frequency $= 1/20 \times 10^{-3}$ s $= 50$ Hz	(1)	3
8 (c)	Half of the wave missing	(1)	
	but with the same frequency as before	(1)	
	(as in diagram, or inverted)		2
			Total 7

Glossary

accurate	when a reading is very close to the true value
alpha particle	a particle consisting of two neutrons and two protons tightly bound together, as in a helium nucleus; emitted as strongly ionising short-range radiation by some unstable isotopes
alternating current (a.c.)	current produced when charges oscillate back and forth, rather than drifting in one direction
annihilation	when particles of matter and antimatter, such as an electron and a positron, collide, are destroyed and their mass is converted to energy
antimatter	the term used to describe matter composed of antiparticles; it is created by high-energy collisions or decays but exists only in small amounts in the observable Universe
antiparticle	a particle of identical mass to a more common particle, but which has opposite values of charge, baryon number and strangeness; denoted by a bar over the symbol, e.g. $\bar{p}$ is an antiproton
atomic number	the number of protons in a nucleus, also called the proton number; has the symbol Z
baryon	any particle composed of three quarks, e.g. a proton
baryon number	a number, denoted by the symbol B, assigned to a particle as a consequence of its quark structure: baryons have $B = 1$, antibaryons have $B = -1$ and all other particles have $B = 0$
beta-minus particle	a fast-moving electron emitted from the nucleus of some unstable isotopes as a result of a neutron decaying into a proton
beta-plus particle	a fast-moving positron emitted from the nucleus of some unstable isotopes as a result of a proton decaying into a neutron
black-body radiator	a body that absorbs all the radiation incident upon it and reflects none, i.e. it is a perfect absorber *and also* a perfect emitter; the surface temperature determines how much energy it emits at each wavelength
boson (or **gauge boson**)	a particle, also called an exchange particle, that 'carries' the force between two particles; for example the photon is the boson that carries the electromagnetic force
cathode	the negative terminal of a power supply, or a negative electrode
charge–mass ratio	the charge carried by a particle, divided by its mass; also called the specific charge; unit $C\,kg^{-1}$
conductor	a material that allows an electric current to pass through it; it has low electrical resistivity
conservation laws	certain physical quantities remain the same for any closed system, and these are said to be conserved quantities; during particle interactions, charge, baryon number and strangeness are conserved: the total value of each of these quantities is the same before and after any interaction

control variable	a variable that is not of interest in an experiment but which may have an impact on the results and so needs to be controlled (fixed)
conventional current	the direction of electric current is taken to be the way that positive charge moves: this is known as conventional current; in metallic conductors it is in fact negative charges (electrons) that move, in the opposite direction to the conventional current
critical temperature	the temperature below which a material becomes superconducting and its electrical resistance drops to zero
current	electric current, I, is the charge, Q, which passes a point in a given time interval, Δt: $I = \Delta Q/\Delta t$; unit A
de Broglie wavelength	the wavelength, λ, of a particle when it behaves like a wave; it depends on the particle's momentum, p: $\lambda = h/p$ where h is the Planck constant
dependent variable	the variable that changes when the independent variable in an experiment is changed
diode	a semiconductor device that allows current to flow only in one direction
direct current (d.c.)	current produced when charges drift in a steady direction
elastic scattering	when particles collide elastically, i.e. when the total kinetic energy of the particles involved in a collision is conserved
electromagnetic wave	a wave that propagates by transferring energy between electric and magnetic fields; does not need a medium in which to travel; light is an electromagnetic wave
electromotive force (e.m.f)	the energy transferred by a power source to each coulomb of charge; equal to the potential difference at the terminals of the source when no current is flowing; unit V
electron	a fundamental particle, a member of the lepton family; carries a negative charge of 1.6×10^{-19} C and has a mass of 9.11×10^{-31} kg
electron volt (eV)	a unit of energy equal to the energy transferred when an electron moves through a potential difference of 1 V
energy	the ability to do work, where work is defined as a force moving through a distance, for example lifting a weight
energy level	specific allowed energy values of electrons in an atom
exchange particle	a particle that mediates the force between two particles; also called a gauge boson
excitation	the process of raising to a higher energy level, for example by collision with a free electron or by the absorption of a photon
femtometre (fm)	unit of length, typical of nuclear dimensions, equal to 10^{-15} m
Feynman diagram	a representation of the exchange of particles in an interaction
fluorescence	the emission of light when excited atoms drop to a lower energy level
forward biased	a diode connected so that the potential difference across it allows it to conduct
frequency	the number of oscillations or waves in one second; unit hertz, Hz

gauge boson	a particle, also called an exchange particle, that 'carries' the force between two particles; for example the photon is the gauge boson that carries the electromagnetic force
gluon	the gauge boson that carries the strong nuclear force
graviton	the gauge boson that carries the force of gravity
ground state	an atom in its ground state has all its electrons in their lowest possible energy level; it cannot emit any photons
hadron	a particle composed of quarks; hadrons are divided into two subgroups, mesons and baryons
hypothesis	a tentative (provisional) idea or theory to explain an observation
independent variable	the variable that is deliberately altered by an experimenter
inelastic scattering	when particles collide and there is a loss of kinetic energy; if an electron is scattered inelastically by an atom, the kinetic energy it loses excites or ionises the atom; inelastic scattering may be followed by photon emission from the excited atom
insulator	a material with hardly any free electrons, which therefore has a very high electrical resistance
internal resistance	the intrinsic electrical resistance between the terminals of a power supply; some energy is always lost inside the supply due to this internal resistance
ionisation	when an atomic electron gains enough energy to escape from the atom; ionisation may follow a collision with a free electron or the absorption of a photon
isotope	isotopes are different forms of atoms of the same element that have the same number of protons and electrons but a different number of neutrons; they are chemically identical but have a different mass number
K meson (kaon)	a meson with unusual decay and interaction properties that carries the conserved quantity, strangeness; there are four types: K^+ composed of an up quark and an antistrange quark K^- composed of an antiup and a strange quark K^0 composed of a down and an antistrange quark $\bar{K}^0$ composed of an antidown and a strange quark
lepton	one of a family of fundamental particles; the electron, muon, tau-particle and neutrino are all leptons
line spectrum	the light from a low-pressure gas, which is emitted at a series of discrete wavelengths, seen through a diffraction grating as sharp lines of different colour; each element has its own characteristic line spectrum
longitudinal wave	a wave in which the oscillations are parallel to the direction of propagation (travel) of the wave; sound travels as a longitudinal wave
meson	a type of hadron formed from a quark–antiquark pair
muon	a fundamental particle in the lepton family that carries the same charge as the electron, but is about 200 times more massive
nanometre (nm)	unit of length, useful for wavelengths of visible light, equal to 10^{-9} m

neutrino	a particle in the lepton family that has a very small mass and no charge; there are three types, the electron-neutrino, the muon-neutrino and the tau-neutrino; they interact only very weakly with other matter
neutron	a hadron of zero charge found in the nucleus of all atoms (except hydrogen-1); free neutrons are unstable and decay to protons via beta-minus decay
neutron number	the number of neutrons, N, in a nucleus; $N = A - Z$
nucleon	any particle (a proton or a neutron) that exists in the atomic nucleus
nucleon number	the total number of protons and neutrons in a nucleus; also referred to as the mass number; symbol A
nucleus	the positively charged, dense matter at the centre of every atom; formed from a combination of neutrons and protons
Ohm's Law	the current, I, through a material is proportional to the potential difference, V, across it; this holds only for certain conductors in certain conditions
ohmic conductor	a material that follows Ohm's Law, e.g. a metal at constant temperature
oscilloscope	a laboratory instrument (a very high resistance voltmeter) which shows on a screen how the potential difference between two points changes with time; can be used to measure the frequency and peak voltage of a.c. signals
pair production	the creation of a particle and its antiparticle from energy; the opposite process to annihilation
peak value	the maximum value (current or potential difference) of an a.c. signal, usually measured from zero; peak voltage is written as V_0, peak current as I_0
peak-to-peak value	the difference between the maximum positive and maximum negative values (current or voltage) of an a.c. signal; for a symmetrical wave the peak-to-peak value is twice the peak value
period	the time taken, T, for one complete oscillation or wave; the reciprocal of the frequency, f: $T = 1/f$
photoelectric effect	the emission of electrons from a metal surface caused by illuminating the metal with electromagnetic radiation
photon	1) a quantum of electromagnetic energy; the energy, E, of a photon is give by $E = hf$, where h is the Planck constant and f is the frequency of the electromagnetic radiation 2) the boson that carries the electromagnetic interaction
pi meson (pion)	a meson (a hadron formed from a quark–antiquark pair); there are three types: π^+ composed of an up quark and an antidown quark π^- composed of an antiup quark and a down quark π^0 composed of an up quark and an antiup quark, or a down quark and an antidown quark
positron	the electron's antiparticle; it has the same mass as an electron and carries an equal but opposite (positive) charge
potential difference (p.d.)	the energy transferred per unit charge (1 C) moving between two points; unit V

potential divider	an arrangement of resistors in series so that the potential difference across the combination is divided between them in the ratio of the resistors
power	the rate at which energy, E, is transferred: power $P = \Delta E/\Delta t$; sometimes expressed as work done per second; unit watt $W = J\,s^{-1}$
precise	when a reading can reliably be given to several significant figures
prediction	a forecast (from a hypothesis or theory) that can be tested by experiment
proton	a positively charged hadron composed of three quarks: up, up and down; it is believed to be the only stable hadron
proton number	the number of protons in a nucleus, also called the atomic number; has the symbol Z
quantum	a discrete amount of a physical quantity, such as energy or charge; for example, charge cannot take any value but has to be a multiple of the charge carried by an electron: charge is said to be quantised
quark	a fundamental particle that is not observed in isolation but always in a combination of three (to make a baryon) or two (a quark–antiquark pair to make a meson); quarks have a baryon number of 1/3 and they carry fractional charge ($\pm2/3$ or $\pm1/3$ of the charge of the electron)
resistance	a measure of how difficult it is for electric current to pass through an object; it is the ratio of potential difference, V, to current, I: resistance $R = V/I$; unit ohm, Ω
resistivity	a property of a material that describes how difficult it is for current to pass through it; for a conducting wire it is related to the resistance, R, the cross-sectional area, A, and the length, l, by the formula: resistivity $\rho = RA/l$; unit Ω m
rest energy	the energy of a stationary particle, E_0, when it is measured by an observer in the same frame of reference; it is linked to the rest mass, m_0, by the equation $E_0 = m_0 c^2$
rest mass	the mass of a stationary particle, m_0, when measured by an observer in the same frame of reference
reverse biased	a diode connected so that the potential difference across it does not allow it to conduct
root mean square (r.m.s.) value	the value of an a.c. current or potential difference that is equal to the d.c. value that would lead to the same power being dissipated in a resistor; the r.m.s. values of I and V are linked to the peak values; $I_{rms} = I_0/\sqrt{2}$ and $V_{rms} = V_0/\sqrt{2}$
specific charge	the charge–mass ratio of a particle; unit $C\,kg^{-1}$
strangeness	a conserved, quantised quantity carried by the strange quark, which has strangeness $S = -1$
strong interaction	the strong nuclear force which acts between quarks and so acts between all hadrons; it has a short range of approximately 10^{-15} m
superconductor	a conductor that has zero resistance at a temperature below its critical value
tau	the heaviest member of the lepton family; it carries an identical charge to the electron but is 3500 times more massive

thermionic emission	the process by which electrons are released from the surface of a heated metal
thermistor	a semiconductor device whose electrical resistance changes significantly with temperature; used as a heat sensor
threshold frequency	the lowest frequency of electromagnetic radiation that causes photo-emission from the surface of a given metal
time base	the control on an oscilloscope that determines the speed at which the electron beam moves horizontally across the screen
transverse wave	a wave in which the oscillations are at right angles to the direction of propagation (travel) of the wave; waves on a string, surface waves on water and electromagnetic waves are all transverse
virtual particle	the short-lived exchange particle (gauge boson) that mediates interactions between particles
wavelength	the distance between identical points on consecutive waves; symbol λ
wave–particle duality	the phenomenon of particles of matter, e.g. electrons, demonstrating both particle and wave properties
weak interaction	a fundamental force that acts between all particles over a very short range, 10^{-18} m; it is responsible for radioactive beta decay
work function	the minimum energy required to remove an electron from the surface of a metal in the photoelectric effect; symbol ϕ
X-plates	charged plates used to deflect the electron beam in an oscilloscope in the horizontal direction
Y-plates	charged plates used to deflect the electron beam in an oscilloscope in the vertical direction

Index

Notes

Notes

Notes

Notes

Notes

Notes